STEVE JONES

The Language of the Genes

Biology, History and the Evolutionary Future

Revised Edition

Flamingo
An Imprint of HarperCollinsPublishers

Flamingo
An Imprint of HarperCollins*Publishers*
77–85 Fulham Palace Road,
Hammersmith, London W6 8JB

www.fireandwater.com

Fully revised edition published by Flamingo 2000
9 8 7 6 5 4 3 2

First published in paperback by Flamingo
with amendments and supplementary bibliographic essay, 1994
Reprinted nine times

First published in hardback in Great Britain by
HarperCollinsPublishers 1993

Author photograph by Sally Soames

ISBN 0 00 655243 9

Set in Linotype Sabon by
Rowland Phototypesetting Ltd, Bury St Edmunds, Suffolk

Printed and bound in Great Britain by
Clays Ltd, St Ives plc

Steve Jones

is Professor of Genetics at the Galton Laboratory, University College London. The first edition of this book was based on the Reith Lectures he gave in 1991. *The Language of the Genes* won both the Rhône-Poulenc prize for the best science book of 1994 and the *Yorkshire Post* Prize for Best First Work. In 1996, the Royal Society presented him with the Michael Faraday Award, given annually to the scientist who has done the most to further the public understanding of science. Professor Jones was born in Wales, educated in Scotland and lives in London. He is author of two other books. *In the Blood* (1996) and *Almost like a Whale* (1999), co-editor of the *Cambridge Encyclopaedia of Human Evolution* and joint author of *Genetics for Beginners* and of the Open University's final-year genetics textbook. On balance he prefers snails to humans.

from the reviews:

'A swashbuckling romp though a subject that normally delights only in its technical opacity.'

RICHARD HORTON, *Literary Review*

'In literate and highly readable style Jones explains genetics and uses it to take us on a tour of human existence.'

A.C. GRAYLING, *Financial Times*

'An inspired grand tour of what genetics can and cannot tell us about ourselves and our evolutionary history. . .wittily written.'

DAVID CONCAR, *New Scientist*

'This urbane, intelligent and informative book deserves to be read widely, since it provides a vivid demonstration of what a remarkable and fragile product of evolution our species is.'

CHRISTOPHER WILLS, *Nature*

'Very witty, very accessible and very smart'

Bite

'A witty tale of curious mutations, molecular clocks, and genetic bottlenecks; it illustrates biological principles with memorable examples from everyday life.'

DOROTHY BONN, *Lancet*

'In an age of genetic triumphalism, Jones is that rare thing – a geneticist who is modest about his subject...Though ultimately optimistic about the potential of modern genetics, the book disarms sceptics by repeatedly warning of the dangers of over-selling molecular biology as a cure-all for human ills.'

JOHN DURRANT, *Observer*

'Jones is one of the best storytellers around today... His thoroughly enjoyable book is scientifically authoritative yet personal, and has a wonderfully dark sense of humour.'

STEVE CONNOR, *Independent on Sunday*

'Jones thoroughly demolishes the old myth that human races are distinct genetic entities, arguing that there is almost as much variation between neighbouring countries as between races... The nature-nurture debate also receives an infusion of sense, this time with the help of the Siamese cat. Ask Jones whether the cat's coat is the work of nature or nurture and you are likely to be told, with more than a hint of exasperation, that the question is meaningless.'

STEPHEN YOUNG, *Guardian*

'Few scientists write well for a general audience, but Steve Jones is exceptional.'

BEVERLEY ANDERSON, Books of the Year, *Observer*

'Good science for thinking people, wide-ranging and informative.' A.S. BYATT, Books of the Year, *Daily Telegraph*

To my parents and my brother
who share my genes and my affection

CONTENTS

Preface

A MALACOLOGIST'S APOLOGY

I have spent – some might say wasted – most of my scientific career working on snails. A malacologist may seem an unlikely author for a book about human genetics. However, my research, when I was still able to do it, was not driven by a deep interest in molluscs. Indeed, one of the few occasions when I thought of giving up biology as a career was when I first had to dissect one. Thirty years ago snails were among the few creatures whose genes could be used to study evolution. They carry a statement of ancestry on their shells in the form of inherited patterns of colour and banding. By counting genes in different places and trying to relate them to the environment one could get an idea of how and why snail populations diverged from each other: of why and how they evolved.

At the time, the idea that it might ever be possible to do the same with humans seemed absurd. Genetics textbooks of the 1960s were routine things. They dealt with the inheritance of pea shape, the sex lives of fungi and the new discoveries about the molecular biology of viruses and their bacterial hosts. Of ourselves, there was scarcely a mention – usually just a short chapter tagged on at the end which described pedigrees of abnormalities such as haemophilia or colour blindness.

Part of this reticence was due to ignorance but part came from the dismal history of the subject. In its early days, the study of human inheritance was the haunt of charlatans, most of whom had a political axe to grind. Absurd pedigrees purporting to show family lines of criminality or genius were the norm. Ignorance and confidence went together. Many biologists argued that it was possible to improve humankind by selective breeding or by the elimination of the unfit. The adulteration of the science reached its disastrous end in the Nazi experiment, and for many years it was seen as at best unfashionable to discuss the nature of inborn differences among people.

After the Second World War, the United Nations published a book – *Statement on Race*, by the American anthropologist Ashley Montagu – which tried to kill some of the genetical myths. I read this as a schoolboy and found it unpersuasive and hard to follow, although its liberal message was clear enough. Re-reading it a few years ago showed why: Ashley Montagu had tried, nobly, to make bricks without straw. The information needed to understand ourselves was simply not available and there seemed little prospect that it ever would be. Human genetics had moved from a series of malign to an equivalent set of pious opinions.

Now everything has been transformed. *Homo sapiens* is no longer the great unknown of the genetical world but has become its workhorse. More is known about the geographical patterns of genes in people than about those of any other animal (snails, incidentally, still come second). The three thousand million letters in the DNA alphabet have, at last, been read from end to end and, so it seems, the century of genetics that began with the rediscovery of Mendel's laws has ended with a new and revolutionary insight into ourselves.

The completion of the DNA map marks the triumph of genetics as a science. Its success as a technology – or, at least, as a medical technology – has yet to be established. Everyone, in the end, dies; and genes are nearly always involved in that unpleasant process. Nobody escapes the fate coded into the double helix. Much of the damage arises anew, either in body cells or as a result of errors in parental sperm and egg. Indeed, most pregnancies end because of such errors. Science has given the hope of finding those at risk of inherited disease and, perhaps, of treating it. At last we understand what sex really means, why we age and die, and how nature and nurture combine to make us what we are.

Most of all, biology has altered our view of our place in the universe of life. For the first time, it is clear how humans are related to other animals and when they first appeared. The idea that Man did not evolve is open to scientific examination: and although creationism is supported by millions the test proves it wrong. Most people believe that they descend from simpler predecessors but would be hard put to say why. As Thomas Henry Huxley, Darwin's great protagonist, said of the idea of evolution: 'It is the customary fate of new truths to begin as heresies and to end as superstitions.' Genetics has saved Darwinism from that fate. It has killed many old and disreputable superstitions. At last there is a real insight into race, and the ancient idea that the peoples of the world are divided into distinct units has gone for ever. Separatism has gained a new popularity among groups anxious to assert an identity of their own, but they cannot call on genes to support their views.

It is, though, the essence of scientific theories that they cannot resolve everything. Science cannot answer the questions that philosophers – or children – ask: why are we here, what is the point of being alive, how ought we to behave? Genetics has nothing to say about what makes people more than just machines driven by biology, about what makes us human. These questions may be interesting, but a scientist is no more qualified to comment on them than is anyone else. Human genetics has suffered from its high opinion of itself. For most of its history it failed to understand its own limits. Knowledge has brought humility to genetics, but its new awareness raises social and ethical problems that have as yet scarcely been addressed.

This book is about what genetics can – and cannot – tell us about ourselves. Its title, *The Language of the Genes*, points to the analogy upon which it turns, the parallels between biological evolution and the history of language.

Inheritance is a discourse through time, a set of instructions passed from generation to generation. It has a vocabulary – the genes themselves – a grammar, the way in which the information is arranged, and a literature, the thousands of instructions needed to make a human being. It is based on the DNA molecule, the famous double helix, the icon of the twentieth century. Johann Miescher, the Swiss discoverer of that marvellous substance, himself wrote in 1892 that its message might be transmitted 'just as the words and concepts of all tongues can find expression in twenty-four to thirty letters of the alphabet.' A century of science shows how right he was.

Both languages and genes evolve. Each generation makes errors in transmission and, sooner or later, enough differences accumulate to produce a new dialect – or a new form of life. Just as the living tongues of the world and their literary relics reveal their extinct ancestors, genes and fossils are an insight into the biological past. We have learned to read the language of the genes and it is saying remarkable things about our history, our present condition and even our future.

The first edition of this book emerged from my Reith Lectures, given on BBC Radio in the early 1990s. Those lectures began with the philosopher Bertrand Russell in 1948 (and, some argue, have gone downhill ever since). I would not dream of comparing myself with my illustrious predecessors but I hope that the series – and the book – can stand on the merits of their subject, the most fascinating in modern science. Perhaps my lectures in their small way helped to show that the BBC can still fulfil its obligations, set forth by its founder Lord Reith, to instruct, inform and entertain. The last might seem an unexpected word to use about science, but it is justified by the number of eccentrics and fools who have graced and disgraced the history of human genetics. They appear sporadically in these pages in the hope of enlivening an otherwise bald narrative.

Since that first edition, seven years ago, genetics – and public concern – have each exploded. What was then remote is here today. In spite of the complaints of Prince Charles, millions of acres of genetically modified crops have been planted; and Dolly the Cloned Sheep, with her penchant for standing on a trough and bullying her inferiors, has been joined by many other domestic animals born without benefit of sex. Some contain genes that make human proteins, as a statement of the new free trade in DNA which makes it possible to move genes from any part of the world of life to any other. We have, with the exception of a few footnotes, read the book of human inheritance. In 2000 it was announced that the order of the DNA bases for every one of the genes needed to make a human being had been established. The rest (small scraps of the 'junk' as it is optimistically called) will be read off within a year or so.

Nobody should disparage this work. The impossible has become commonplace. To decipher the DNA has been an enormous task. It was, briefly, the privilege of a professor (or his technicians). Then came the time of the postgraduates, with doctorate after doctorate awarded for one or other piece of the genetic jigsaw. Soon, the machines took over – cheaper, less subject to emotional upset, and far faster than even the most dedicated student. Brute force (helped by ingenuity) triumphed and the pace of discovery accelerated in a fashion more associated with computers than with biology. Part of the rush came from the excitement of a science armed with a goal and the technology to reach it, but part emerged from an attempt to make millions from patents and a competing effort to keep the information in the public domain.

The need for funds and the prospect of fortune has given birth to an era of exaggerated hopes and fears about inheritance. The public is obsessed with genes. In part that is because they come close to questions that lie outside science altogether; issues of sex, identity and fate that have occupied sages since the days of the Old Testament, the first genetics text of all. Genetics is more and more involved with social and political questions such as those of abortion, cloning, and human rights. It puts medical issues into sharp and sometimes uncomfortable focus, with much concern about problems of privacy, blame and the nature of disease. Many inherited illnesses are expensive to treat and hard to cure. They raise unwelcome questions about the balance of responsibility between individuals and populations.

Much has been spent in the past decade. Those who paid for the map of the genes are anxious for some return. It can be hard to translate theory into practice. Vesalius worked out the anatomy of the heart in 1543; but the first heart transplant was not until 1967. Although it will not take as long before gene transplants arrive, they are further away than most people realise, and one important task that genetics faces (and one of the aims of this book) is to tailor public demands to reality.

The new genetics sounds (and is) both beguiling and alarming. Some of those involved have been quick to take advantage of public naiveté and have maintained a stream of promises as to what they will soon achieve. Few have been fulfilled; and some will not be. The molecular biology business promotes its wares as well as any other, and the four letters of the genetic code might nowadays well be restated as H, Y, P and E. Even so, in genetics, more than most sciences, fantasy has a habit of turning into reality in unpredictable ways, even as much-heralded break-throughs do not appear. At the time of my first edition, the idea that inherited disease would be cured with gene therapy was just around the corner, where it remains. At that time, though, the idea that animals – perhaps even humans – might routinely be cloned, or that lengths of DNA could be moved around at will seemed beyond belief. Now, genetic engineering is a business worth billions a year.

The biggest change in the past seven years has been in attitude. In the public mind, genetics is no longer a science but a faith; a curse or a salvation. It promises or threatens, according to taste. In fact, biology has told us little about human affairs that we did not know before. Both have had plenty of publicity. Dozens of works of exegesis now offer salvation in a molecular paradise or (choose your Church) eternal damnation to those who take the broad path down the double helix to Hell. Some are accounts of the latest advances, but too many are in that weary penumbra of science inhabited by sociologists, who wander like children in a toyshop, playing with devices they scarcely under-stand. Biochemistry has become a branch of the social sciences and, some say, life will be explained in genetic terms. Many welcome the idea, some are filled with horror, but few pause to consider what, if anything, it means.

The public needs a fairer statement of what science can and cannot do. Reality is harder to sell than hopes or fears; but DNA deserves more than the Frankensteins and designer babies that fill the press. The problem is, at all levels, one of unreasonable expectation, both positive and negative. In this revised version of *The Language of the*

Genes I try to cover the many advances since its first version; in the map of human DNA, in the genetic manipulation of plants and animals, and in our new abilities to screen for inborn disease. I have tried to keep the book to size and have thrown out several sections to allow space for the developments of the past decade.

Since this work first appeared, my malacological career has taken second place to journalism. Perhaps, in time, human genetics will help to understand the world of snails, so that this episode of reporting, rather than doing, will not be wasted.

JSJ, June 2000.

Introduction

THE FINGERPRINTS OF HISTORY

In 1902, in Paris, a horrible murder was solved by the great French detective Alfonse Bertillon. He used a piece of new technology which struck fear into the heart of the criminal community. Eight decades later two young girls were killed near the Leicestershire village of Narborough. Again, the murderer was found through a technical advance, although the machinery involved would have been beyond the comprehension of Bertillon. These events link the birth and the coming-of-age of human genetics.

The Parisian killer was trapped because he left a fingerprint at the scene of the crime. For the first time, this was used in evidence as a statement of identity. The idea came from ancient Japan, where a finger pressed into a clay pot identified its maker. The Leicestershire murderer was caught in the same way. A new test looked for individual differences in genetic material. This 'DNA fingerprint' was as much a statement of personal uniqueness as Bertillon's clue or the potter's mark. As usual, life was more complex than science. The killer, a baker called Colin Pitchfork, was caught only after DNA fingerprints had eliminated a young man who had made a false confession and after Pitchfork had persuaded a friend to give a fraudulent blood sample under his name.

The idea that fingerprints could be used to trace criminals came from Charles Darwin's cousin, Francis Galton. He founded the laboratory in which I work at University College London, the first human genetics institute in the world. Every day I walk past a collection of relics of his life. They include some rows of seeds that show similarities between parents and offspring, an old copy of *The Times* and a brass counting gadget that can be hidden in the palm of the hand. Each is a reminder of Galton. As well as his revolution in detective work Galton was the first person to publish a weather map and the only one to have made a beauty map of Britain, based on a secret ranking of the local women on a scale of one to five (the low point, incidentally, being in Aberdeen).

His biography reveals an unrelieved eccentricity, well illustrated by the titles of a dozen of his three hundred scientific papers: On spectacles for divers; Statistical inquiries into the efficacy of prayer; Nuts and men; The average flush of excitement; Visions of sane persons; Pedigree moths; Arithmetic by smell; Three generations of lunatic cats; Strawberry cure for gout; Cutting a round cake on scientific principles; Good and bad temper in English families; and The relative sensitivity of men and women at the nape of the neck. Galton travelled much in Africa, regarding the natives with some contempt and measuring the buttocks of the women using a sextant and the principles of surveying.

Galton's work led, indirectly, to today's explosion in human genetics. His particular interest was in the inheritance of genius (a class within which he placed himself). In his 1869 book *Hereditary Genius,* he investigated the ancestry of distinguished people and found a tendency for talent to crop up again and again in the same family. This, he suggested, showed that ability was inborn and not acquired. *Hereditary Genius* marked the first attempt to establish patterns of human inheritance with well-defined traits – such as becoming (or failing to become) a judge – rather than with mere speculation about vague qualities such as fecklessness.

Galton and his followers would be astonished at what biology can now do. It still does not understand attributes such as genius (and reputable scientists hardly concern themselves with them), but DNA is much involved in mental and physical illness. Half a million DNA samples have been taken by police in Britain since the test was invented, and the government has a scheme to follow the genes – and the ailments – of the same number of its citizens over two decades in the hope of finding the biological errors responsible for killers like cancer and heart disease. New tests mean that parents can sometimes choose whether to risk the birth of a child with an inborn defect. Ten thousand such illnesses are known and if we include, as we should, all ailments with an inherited component, most people die because of the genes they carry.

Genetics does more than reveal fate. Humans share much of their heritage with other creatures. As Galton himself illustrated with the appropriate impression pasted near that made by Gladstone, the prime minister, chimpanzees have fingerprints. Now we know that much of their DNA is identical to our own (as indeed is that of bananas). All this suggests that humans and apes are close relatives.

Genetics is the key to the past. As every gene must have an ancestor, inherited diversity can be used to piece together a picture of history more complete than from any other source. Each segment of DNA is a message from our forebears and together they contain the whole story of human evolution. Everyone alive today is a living fossil and carries within themselves a record that revisits the birth of humankind. *The Origin of Species* expresses the hope that 'light will be thrown on the origin of man and his history'. Darwin's hint that humans share a common descent with all other creatures is now accepted by all scientists, because of the evidence of the genes.

Evolution, the appearance of new forms by the alteration of those already present, is no more than descent with modification. The same is true of language. As a boy, I was amused by the tale of the order going down the line of command to soldiers in the trenches. 'Send reinforcements, we're going to advance' changed to 'Send three and fourpence, we're going to a dance' as it passed from man to man. This simple tale illustrates how accidents, as an inherited message is copied, can lead to change. Because of mutation, life, too, is garbled during transmission.

This book is about inheritance: about the clues of our past, present and future that we all contain. The language of the genes has a simple alphabet, with not twenty-six letters, but four; the DNA bases – adenine, guanine, cytosine and thymine (A, G, C and T for short). They are arranged in words of three letters such as CGA or TGG. Most code for different amino acids, which are themselves joined together to make proteins, the building blocks of the body.

The economy of life's language can be illustrated with an odd quotation from a book called *Gadsby*, written in 1939 by one Ernest Wright: 'I am going to show you how a bunch of bright young folks did find a champion, a man with boys and girls of his own, a man of so dominating and happy individuality that youth was drawn to him as a fly to a sugar bowl.' This sounds somewhat peculiar, as does the rest of the fifty-thousand word book, and it is. The quotation, and the whole work, lacks the letter 'e'. An English sentence can be written with twenty-five letters instead of twenty-six, but only just. Biology manages with a mere four.

Although its vocabulary is simple the genetical message is very long. Each cell in the body contains about six feet of DNA. There are so many cells that if all the DNA in a single human body were stretched out it would reach to the moon and back eight thousand times. Twenty years

4

ago, the Human Genome Project set out to read the whole
of its three thousand million letters, and to publish perhaps
the most dreary volume ever written, the equivalent of a
dozen or so copies of the *Encyclopaedia Britannica*. The
task is now more or less complete. The sequencers followed
a grand scientific tradition: the Admiralty, after all, sent
the *Beagle* to South America with Darwin on board not
because they were interested in evolution but because they
knew that if they were to understand (and, with luck,
control) the world, the first step was to map it. The chart
of the genes, like that of the Americas, has been expensive
to make; but – like the theory of evolution itself – it may
change our perception of ourselves.

Powerful ideas like inheritance and evolution soon
attract myths. Impressed by his studies of genius, Galton
founded the science (if that is the right word) of eugenics.
Its main aim was 'to check the birth rate of the Unfit and
improve the race by furthering the productivity of the fit
by early marriage of the best stock'. He led the new field
of human genetics into a blind alley from which it did not
emerge for half a century. At his death, he left £45,000 to
found the Laboratory of National Eugenics at University
College London and, in fine Victorian tradition, £200 to
his servant who had worked for him for forty years. His
research institute soon changed its name to the Galton
Laboratory to escape from the eugenical taint. What
became of his servant is not recorded.

Galton's social ideas and Darwin's evolutionary insights
had a pervasive effect on the intellectual history of the
twentieth century. They influenced left and right, liberal
and reactionary, and continue – explicitly or otherwise –
to do so. Many disparate figures trace their ideas to *The
Origin* and to *Hereditary Genius*. All are united by one
belief: in biology as destiny, in the power of genes over
those who bear them.

The most famous monument in Highgate Cemetery in London, a couple of miles north of today's Galton Laboratory, is that of Karl Marx. Its inscription is well known: 'Philosophers have only interpreted the world. The point, however, is to change it.' Darwinism was soon used in an attempt to live up to that demand. The philosopher Herbert Spencer, buried just across the path from Marx, founded what he called Social Darwinism; the notion that poverty and wealth are inevitable as they reflect the biological rules that govern society. In his day, Spencer was famous. His *Times* obituary claimed that 'England has lost the most widely celebrated and influential of her sons.' Now he is remembered only for that neatly circular phrase 'the survival of the fittest' and for inventing the word 'evolution'.

He wrote with a true philosopher's clarity: 'Evolution is an integration of matter and concomitant dissipation of motion; during which matter passes from an indefinite, incoherent homogeneity, to a definite, coherent heterogeneity, through continuous differentiations and integrations'. Those lucid lines were parodied by a mathematical contemporary: 'A change from a nohowish, untalkaboutable all-alikeness to a somehowish and in general talkaboutable not-all-alikeness by continuous somethingelsifications and sticktogetherations.'

Spencer used *The Origin of Species* as a rationale for the excesses of capitalism. The steel magnate Andrew Carnegie was one of many to be impressed by the idea that evolution excuses injustice. He invited Herbert Spencer to Pittsburgh. Unfortunately, the philosopher's response to his trip to see his theories worked out in steel and concrete was that 'Six months' residence here would justify suicide.'

Galton, too, supported the idea of breeding from the best and sterilising those whose inheritance did not meet with his approval. The eugenics movement joined a gentle concern for the unborn with a brutal rejection of the rights of the living (a combination not unknown today). Galton's main interest in genetics was as a means to forestall the imminent decline of the human race. He claimed that families of 'genius' had fewer children than most and was concerned about what this meant for the future. It was man's duty to interfere with his own evolution. As he said: 'What Nature does blindly and ruthlessly, man may do providently, quickly and kindly.' Perhaps his own childless state helped to explain his anxiety.

Many of the eugenicists shared the highly heritable attributes of wealth, education and social position. Francis Galton gained his affluence from his family of Quaker gunmakers. Much of his agenda was the survival of the richest. Other eugenicists were on the left. They felt that if economies could be planned, so could genes. George Bernard Shaw, at a meeting attended by Galton in his last years, claimed that 'Men and women select their wives and husbands far less carefully than they select their cashiers and cooks.' Later, he wrote that 'Extermination must be put on a scientific basis if it is ever to be carried out humanely and apologetically as well as thoroughly.' Shaw was, no doubt, playing his role as Bad Boy to the Gentry, but subsequent events made his tomfoolery seem even less droll than it did at the time.

Sometimes, such notions were put into practice. Paraguay has an isolated village with an unusual name: Nueva Germania, New Germany. Many of its inhabitants have blonde hair and blue eyes. Their names are not Spanish, but are more likely to be Schutte or Neumann. They are the descendants of an experiment; an attempt to improve humankind. Their ancestors were chosen from the people of Saxony in 1886 by Elisabeth Nietzsche – sister of the philosopher, who himself uttered the immortal phrase 'What in the world has caused more damage than the follies of the compassionate?' – as particularly splendid specimens, selected for the purity of their blood. The idea was suggested by Wagner (who once planned to visit). The New Germans were expected to found a community so favoured in its genetic endowment that it would be the seed of a new race of supermen. Elisabeth Nietzsche died in 1935 and Hitler himself wept at her funeral. Today the people of Nueva Germania are poor, inbred and diseased. Their Utopia has failed.

The eugenics movement had an influence elsewhere in the New World. In 1898, Charles Davenport, then professor of evolutionary biology at Harvard, was appointed as Director of the Cold Spring Harbor Laboratory on Long Island Sound. Initially, the Laboratory concentrated on the study of 'the normal variation of the animals in the harbor, lakes and woods, and the production of abnormalities'. It carried out some of the most important work in early twentieth-century biology.

Soon, Mrs E H Harriman, widow of the railway millionaire, decided to devote part of her fortune to the study of human improvement. The Eugenics Record Office was built next to the original laboratory. It employed two hundred field workers, who were sent out to collect pedigrees. Their 750,000 genetic records included studies of inherited disease and of colour blindness; but also recorded the inheritance of shyness, pauperism, nomadism, and moral control.

Davenport's work had an important effect on American society. The first years of the twentieth century saw eugenical clubs with prizes for the fittest families and, for the first time, medicine became concerned about whether its duty to the future outweighed the interests of some of those alive today. In Galtonian style, Davenport claimed that: 'Society must protect itself; as it claims the right to deprive the murderer of his life so also may it annihilate the hideous serpent of hopelessly vicious protoplasm.' Twenty-five thousand Americans were sterilised because they might pass feeble-mindedness or criminality to future generations. One judge compared sterilisation with vaccination. The common good, he said, overrode individual rights.

Another political leader had similar views. 'The unnatural and increasingly rapid growth of the feeble-minded and insane classes, coupled as it is with steady restriction among all the thrifty, energetic and superior stocks constitutes a national and race danger which it is impossible to exaggerate. I feel that the source from which the stream of madness is fed should be cut off and sealed off before another year has passed.' Such were the words of Winston Churchill when Home Secretary in 1910. His beliefs were seen as so inflammatory by later British governments that they were not made public until 1992.

One of Galton's followers was the German embryologist Ernst Haeckel. Haeckel was a keen supporter of evolution. He came up with the idea (which later influenced Freud) that every animal re-lived its evolutionary past during its embryonic development. His interest in Galton and Darwin and his belief in inheritance as fate led him to found the Monist League, which had thousands of members before the First World War. It argued for the application of biological rules to society and for the survival of some races – those with the finest heritage – at the expense of others. Haeckel claimed social rules were the natural laws of heredity and adaptation. The evolutionary destiny of the Germans was to overcome inferior peoples: 'The Germans have deviated furthest from the common form of ape-like men ... The lower races are psychologically nearer to the animals than to civilized Europeans. We must, therefore, assign a totally different value to their lives.'

In 1900 the arms manufacturer Krupp offered a large prize for the best essay on 'What can the Theory of

Evolution tell us about Domestic Political Development and the Legislation of the State?' There were sixty entries. In spite of the interests of capital, the first German eugenic sterilisation was carried out by a socialist doctor (albeit one who claimed that trade union leaders were more likely to be blond than were their followers).

While imprisoned after the Beer Hall Putsch, Hitler read the standard German text on human genetics, *The Principles of Human Heredity and Race Hygiene*, by Eugene Fischer. Fischer was the director of the Berlin Institute for Anthropology, Human Heredity and Eugenics. One of his assistants, Joseph Mengele, later achieved a certain notoriety for his attempts to put Galtonian ideas into practice. Fischer's book contained a chilling phrase: 'The question of the quality of our hereditary endowment' – it said – 'is a hundred times more important than the dispute over capitalism or socialism.'

His thoughts were echoed in *Mein Kampf*: 'Whoever is not bodily and spiritually healthy and worthy shall not have the right to pass on his suffering in the body of his children'. Hitler took this to its dreadful conclusion with the murder of those he saw as less favoured in order to breed from the best. The task was taken seriously, with four hundred thousand sterilisations of those deemed unworthy to pass on their genes, sometimes by the secret use of X-rays as the victims filled in forms. Those in charge of the programme in Hamburg estimated that one fifth of its people deserved to be treated in this way.

By 1936 the German Society for Race Hygiene had more than sixty branches and doctorates in racial science were offered at several German universities. Certain peoples were, they claimed, inferior because of inheritance. Half of those at the Wannsee Conference (which decided on the final solution of the Jewish problem) had doctorates and many justified their crimes on scientific grounds. The eugenics movement in Germany was opposed to abortion (except of the unfit) and imposed stiff penalties – up to ten years in prison – on any doctor rash enough to carry it out. The number of children born to women of approved stock went up by a fifth. The Hitlerian conjunction of extreme right wing views, an obsession with racial purity and a hatred of abortion has its echoes today.

Concern for the purity of German blood reached absurd lengths. One unfortunate member of the National Socialist Party received a transfusion from a Jew after he had been in a road accident. He was brought before a disciplinary court to see if he should he excluded from the Party. Fortunately, the donor had fought in the First World War, so that his Jewish red cells were – just about – acceptable.

The disaster of the Nazi experiment ended the eugenics movement, at least in its primitive form. Its blemished past means that human genetics is marked by the fingerprints of its own history. It sometimes seems to find them hard to wipe off. They should not be forgotten now that the subject is, for the first time, in a position to control the biological future.

Galton and his followers felt free to invent a science which accorded with their own prejudices. They believed that the duty to genes outweighs that to those who bear them. They were filled with extraordinary self-assurance and great weight was placed on their views although in retrospect it is obvious that they knew almost nothing.

Today's new knowledge is as controversial as was the old ignorance. Even so, disputes among modern biologists are not about the vague general issues that obsessed their predecessors. Instead they concern themselves with the fate of individuals rather than of all humanity. Genetics has become a science and, as such, has narrowed its horizons.

Nevertheless, it raises ethical issues which will not go away. The newspapers are filled with debates about the morals of gene therapy or of human cloning, neither of which show any sign of becoming a reality. However, the diagnosis of defective genes before birth has already shifted the balance between birth and abortion to reduce the number of damaged children. This raises passions, from those who feel – in spite of the high natural wastage of fertilised eggs – that all foetuses are sacred, to others who consider that to pass on a faulty gene is equivalent to child abuse. Genetics presents a more universal difficulty – the problem of knowledge. Soon, it will tell many of us how and when we may die. Already, it is possible to diagnose at birth genes which will kill in childhood, youth or middle age. More will soon be found. Will people want to know that they are at risk of a disease which cannot be treated? Many genes show their effects in those who inherit damaged DNA from each parent. As everyone is likely to pass on a single copy of at least one such gene, will this help to choose a partner or to decide whether to have children?

Attitudes to inborn disease are flexible. In Ghana, babies are sometimes born with an extra finger or toe. Some tribal groups take no notice, others rejoice as it means that the new member of the family will become rich; but others, just a few miles away, regard such children with horror and they are drowned at birth. Even Christianity has seen the genetically unfortunate as less than human. Martin Luther himself declared that Siamese twins were monsters without a soul. Attitudes to genetics will always be influenced by those to abortion, which vary with time and place. St Augustine saw a foetus as part of its mother and not worthy of protection and in spite of its present views the Catholic Church did not condemn abortion until the thirteenth century. Ireland has a constitutional clause that establishes the right to life of the unborn child; while across the Irish Sea abortion until the third month is available almost on demand. Embryo research (which is becoming important with the discovery that embryonic cells can be used to treat adult disease) is forbidden in Germany but lightly controlled in Britain. All this shows how hard it is to set ethical limits to the new biology.

The problem can be illustrated with some old-fashioned biological discrimination. There has always been prejudice against certain genes, those carried on the chromosomes that determine sex. Women have two 'X' chromosomes, men a single X chromosome and a much smaller 'Y'. All eggs have an X but that of sperm are of two kinds, X or Y. At fertilisation, both XY males and XX females are produced in equal number. Sex is as much a product of genes as are blood groups.

How the value of these genes is judged shows how biological choice can depend on circumstances. Sometimes, Y chromosomes seem to be worth less than Xs. When it comes to wars, murders and executions, males have always been more acceptable victims than females. But the balance can shift. Many parents express a preference for sons, especially as a first-born. Some even try to achieve them. The recipes vary from the heroic to the hopeful. In ancient Greece, to tie off the left testicle was said to do the job, while mediaeval husbands drank wine and lion's blood before copulating under a full moon. Less drastic methods included sex in a north wind and hanging one's underpants on the right side of the bed.

To sell gender is an easy way to make money. It has, after all, a guaranteed fifty per cent success rate. Today's methods vary from the use of baking soda or vinegar at the appropriate moment (to take advantage of a supposed difference in the resistance of X and Y-bearing sperm to acids and alkalis) to sex at particular times of the female cycle. A diet high or low in salt is also said to help. Such recipes are useless and some of those who sell them have been prosecuted for fraud.

Now, fraud is out of date. Sex can be chosen in many ways. One is to separate X and Y sperm and to fertilise a woman with the appropriate type. The methods are not absolute, but shift the ratios by about two to one for males and four to one for females. Since Louise Brown in 1978, thousands of children have been born by in-vitro fertilisation, with sperm added to egg in a test-tube. A single cell can be taken from the embryo and its sex determined (and, indeed, as young male embryos grow faster, simply to choose the largest embryo biases the ratio of males). Only those of the desired gender are implanted into the mother. This technique has led to the birth of hundreds of babies.

Pregnancy termination is a less kind, but equally effective, way of choosing the sex of a child. Aristotle himself felt that a male foetus should be protected from abortion after forty days, but a female only after ninety. A recent survey of geneticists themselves showed that, in Holland, none would accept pregnancy termination just to choose the sex of a child, in Britain one in six, and in Russia nine out of ten. The Indian government was forced to shut down clinics which chose the sex of a baby with a test of the chromosomes of the foetus and aborted those with two Xs. More than two thousand pregnancies a year were ended for this reason in Bombay alone. The main reason was the need for large dowries when daughters were married off. The advertisements said 'Spend six hundred rupees now, save fifty thousand later.' The preference is an old one. A nineteenth-century visitor to Benares recorded that 'Every female infant in the Rajah's family born of a lawful wife, or Rani, was drowned as soon as it was born in a hole in the earth filled with milk.' The rulers' many wives were said to have produced no grown-up daughters for more than a century. The government nowadays pays a bonus for girl babies, but some states now have four females to five males and the country as whole has a deficit of girls and women equivalent to the entire British female population.

All these methods interfere with genes. Their acceptability varies from the reasonably uncontentious choice of sperm to a crime where the murder of girl children is concerned. Where to draw the line depends on one's own social, political or religious background; on how acceptable the notion might be that fate should depend on biological merit. All readers of this book would, I imagine, abhor infanticide, and most might feel that to terminate a pregnancy just because it is the wrong sex was also wrong. They might worry less about the choice of X or Y sperm.

The choice of a child's sex can, however, involve more than parental self-indulgence. Sometimes it is a matter of life and death. Many inherited diseases are carried on the X chromosome. In most girls, an abnormal X is masked by a normal copy. Boys do not have this option, as they have but a single X. For this reason, sex-linked abnormalities, as they are known, are much more common in boys than in girls. They can be distressing. Duchenne muscular dystrophy is a wasting disease of the muscles. Symptoms can appear even in three year-olds and affected children have to wear leg braces by the age of seven, are often in a wheelchair by eleven and may die before the age of twenty-five. Parents who have seen one of their sons die of muscular dystrophy are in the agonising position of knowing that any later son has a one in two chance of having inherited it. A couple who have had a son with the illness can scarcely be blamed for a desire to ensure that no later child is affected. They hope to control the quality of their offspring and few will criticise them for doing so. Genetics has changed their ethical balance.

If a couple has a son with muscular dystrophy they know at once that the mother carries the gene. The chance of a second son with the disease is hence far greater than before. It is still just one in two, so that to terminate all male pregnancies means a real possibility of losing a normal boy. Even those who dislike the idea of choice of a child's sex with X-bearing sperm might change their minds in these circumstances. Others would go further and accept the option of an externally fertilised embryo or the termination of all pregnancies which would produce a son.

Now, such choices have become more precise. The gene for muscular dystrophy has been found and changes in the DNA can show whether a foetus bears it. Hundreds of centres use the test. But the method is far from perfect. The gene can go wrong in many ways and not all of them show up. A foetus that appears normal may hence, in a proportion of cases, carry the gene. This complicates the parents' decision as to whether to continue with a pregnancy. To sample foetal tissues also involves a certain hazard. This has become smaller as technology improves, with a check of foetal cells in the mother's blood, but the risks of the test must themselves be weighed in the moral scales.

As more is found about the genes that cause death not at birth, or in the teens, but in middle or old age the dilemmas increase. Given the opportunity, some might avoid the birth of a baby doomed to dementia through Alzheimer's disease in its forties. Others would argue that forty years of life are not to be dismissed; and that, in four decades of science, the cure may be found.

Decisions about the future of an unborn child will, as a result, more and more be influenced by estimates of risk and of quality: by whether the rights of a foetus depend on its genes. Such judgements are not just scientific decisions, but depend on the society and the people who make them. The debacle of the eugenics movement led to an understandable reluctance even to consider the idea of choices about rights based on inherited merit, but the new knowledge means that they are unavoidable.

Galton himself would have been delighted by the idea of preventing the birth of the damaged. The new eugenics can be overt. The *Chinese People's Daily* is frank in its views. It reported a scheme to ban the marriage of those with mental disease unless they were sterilised with a robust simplification of Mendelism: 'Idiots give birth to idiots!' the eugenical message is often justified on financial grounds. At the Sesquicentennial Exhibition in Philadelphia in 1926 the American Eugenics Society had a board that counted up the $100 per second supposed to be spent on people with 'bad heredity'. Sixty years later, one proponent of the plan to sequence the human genome claimed that the project would pay for itself by 'curing' schizophrenia – by which he meant the termination of pregnancies carrying the as yet hypothetical and undiscovered gene for the disease. The 1930s were a period of financial squeeze for health care. Seventy years on, the state is still anxious to limit the amount spent on medicine in the face of an inexorable rise in costs, with inborn diseases among the most expensive. There is a fresh danger that genetics will be used as an excuse to discriminate against the handicapped in order to save money.

Genetics – science as a whole – owes its success to the fact that it is reductionist: that to understand a problem, it helps to break it down into its component parts. The human genome project marks the extreme application of such a view. The approach works well in biology as far as it goes, but it only goes so far. Its limits are seen in a phrase once notorious in British politics, the late Prime Minister Mrs Thatcher's statement that 'There is no such thing as society, there are only individuals.' The failures of her philosophy are all around us. To say, with Galton and his successors, 'There are no people, there are only genes' is to fall into the same trap.

In spite of the lessons of the past, there has been a

resurgence of the dangerous and antique myth that biology can explain everything. Some have again begun to claim that we are controlled by our inheritance. They promote a kind of biological fatalism. Humanity, they say, is driven by its inheritance. The predicament of those who fail comes from their own weakness and has little to do with the rest of us. Such *nouvelle* Galtonism suggest that human existence is programmed and that, apart from a little selective pregnancy termination, there is no point in any attempt to change it – which is convenient for those who like things the way they are.

After the Second World War, genetics had – it seemed – at last begun to accept its own limits and to escape its confines as the haunt of the obsessed. Most of those in the field today are cautious about claims that the essence of humanity lies in DNA. Although it can say extraordinary things about ourselves, genetics is one of the few sciences that has reduced its expectations.

In mediaeval Japan, the science of dactylomancy – the interpretation of personality from fingerprints – had it that people with complex patterns were good craftsmen, those with many loops lacked perseverance, while those whose fingers carried an arched pattern were crude characters without mercy. Human genetics has escaped from its dactylomantic origins. The more we learn about inheritance the more it seems that there is to know. The shadow of eugenics has not yet disappeared but is fainter than it was. Now that genetics has matured as a subject it is beginning to reveal an extraordinary portrait of who we are, what we were, and what we may become. This book is about what that picture contains.

Chapter One

A MESSAGE FROM OUR ANCESTORS

The rich were the first geneticists. For them, vague statements of inherited importance were not enough. They needed – and awarded themselves – concrete symbols of wealth and consequence that could persist when those who invented them were long dead. The Lion of the Hebrew Tribe of Judah was, until a few years ago, the symbol of the Emperor of Ethiopia, while those of England descend from the lions awarded to Geoffroy Plantagenet in 1177. The fetish for ancestry means that royal families are important in genetics (Prince Charles, for example, has 262,142 ancestors recorded on his pedigree). The obsession persists against all attempts to deny it. Heraldry was cut off by the American Revolution, but George Washington himself attempted to make a connection with the Washingtons of Northamptonshire and used, illegally, their five-pointed stars as a book plate.

Heraldic symbols were invented because only when the past is preserved does it make sense. For much of history wealth was dissipated on funerary ornaments to remind the unborn from whence they sprang. University College London contains an eccentric object; the stuffed body of the philosopher Jeremy Bentham (who was associated with the College at its foundation). Bentham hoped to start a fashion for such 'auto-icons' in the hope of reducing the cost of monuments to the deceased. It did not catch on, although the popularity of his corpse with visitors suggests that it ought to have done.

Such pride in family would now be greeted, mainly, with derision. Harold Wilson, the British Prime Minister of the 1960s, did as much when he mocked his predecessor, Lord Home, for being the Seventeenth Earl of that name. Lord Home deflected the jest when he pointed out that his critic must be the seventeenth Mr Wilson. He made a valid claim: that while only a few preserve their heritage in an ostentatious way, every family, aristocratic or not, retains the record of their ancestors. Everyone, however deficient in history, can decipher their past in the narrative of the DNA.

Some can use inherited abnormalities. A form of juvenile blindness called hereditary glaucoma is found in France. Parish records show that most cases descend from a couple who lived in the village of Wierr-Effroy near Calais in the fifteenth century. Even today pilgrims pray in the village church of Sainte Godeleine, which contains a cistern whose waters are believed to cure blindness. Thirty thousand descendants have been traced and for many the diagnosis of the disease was their first clue about where their ancestors came from and who their relatives might be. The gene went with French emigrants to the New World.

Human genetics was, until recently, restricted to studying pedigrees that stood out because they contained an inborn disease. Its ability to trace descent was limited to those few kindreds who appear to deviate from some perfect form. Biology has now shown that perfection is a mirage and that, instead, variation rules. Thousands of characters – normal diversity, not diseases – distinguish each nation, each family and each person. Everyone alive today is different from everyone who ever has lived or ever will live. Such variation can be used to look at shared ancestry in any lineage, healthy or ill, aristocratic or plebeian. Every modern gene brings clues from parents and grandparents, from the earliest humans a hundred thousand years and more ago and from the origin of life four thousand million years before that.

Most of genetics is no more than a search for diversity. Some differences can be seen with the naked eye. Others need the most sophisticated methods of molecular biology. As a sample of how different each individual is we can glance beneath the way we look to ask about variation in how we sense the world and how the world perceives us.

Obviously, people do not much resemble each other. The inheritance of appearance is not simple. Eye colour depends first on whether any pigment is present. If none is made the eye is pale blue. Other tints vary in the amounts of the pigment made by several distinct genes, so that colour is not a dependable way of working out who fathered a particular child. The inheritance of hair type is also rather complex. Apart from very blonde or very red hair, the genetics of the rest of the range is confused and is further complicated by the effects of age and exposure to the sun.

Even a trivial test shows that individuals differ in other ways. Stick your tongue out. Can you roll it into a tube? About half those of European descent can and half cannot. Clasp your hands together. Which thumb is on top? Again, about half the population folds the left thumb above the right and about half do it the other way. These attributes run in families but their inheritance, like that of physical appearance, is uncertain.

People vary not just in the way the world sees them, but how they see it. A few are colour-blind. They lack a receptor for red, green or blue light. All three are needed to perceive the full range of colour. The absence of (or damage to) one (usually that for green, less often for red, almost never for blue) gives rise to a mild disability that may have made a difference when gathering food in ancient times. The three genes involved have now been tracked down. Those for red and green are similar and diverged not long ago, while the blue receptor has an identity of its own. John Dalton, best known for his atomic theory, was himself so colour-blind as to match red sealing-wax with a leaf (which must have made things difficult for a chemist). He believed that his own eyes were tinted with a blue filter and asked that they be examined after his death. They were, and no filter was found, but, a century and a half later, a check of the DNA in his pickled eyeballs showed him to have lacked the green-sensitive pigment.

Colour-blindness marks the extreme of a system of normal variation in perception. When asked to mix red and green light until they match a standard orange colour, people divide into two groups that differ in the hue of the red light chosen. There are two distinct receptors for red, differing in a single change in the DNA. About sixty per cent of Europeans have one form, forty per cent the other. Both groups are normal (in the sense that they are aware of no handicap) but one sees the world through rather more rose-tinted spectacles than the other. The contrast is small but noticeable. If two men with different red receptors were to choose jacket and trousers for Father Christmas there would be a perceptible clash between upper and lower halves.

In the 1930s, a manufacturer of ice trays was surprised to receive complaints that his trays made ice taste bitter. This baffled the entrepreneur as the ice tasted just like ice to him, but was a hint of inherited differences in the ability to taste. To some, a trace of a substance used in the manufacturing process is intolerable, while to others a concentration a thousand times greater has no taste at all. Much of the difference depends on just one gene which exists in two forms. That observation, the ability or otherwise to perceive a substance, now called PROP, was the key to a new universe of taste. Genetic 'supertasters' are very sensitive to the hops in beer, to pungent vegetables like broccoli, to sugar and to spices, while non-tasters scarcely notice them. Half the population of India cannot taste the chemical at all, but just one African in thirty is unable to perceive it. Students of my day thought it witty to make tea containing PROP to see the bafflement of those who could drink it and those who could not. Today's undergraduates have more sense.

As truffle-hunters know, scent and taste are related. There is genetic variation in the ability to smell, among other things, sweat, musk, hydrogen cyanide and the odour of freesias. Many animals communicate with each other through the nose. Female mice can smell not only who a male is, but how close a relative he might be. Humans also have an odorous identity, as police dogs find it more difficult to separate the trails of identical twins (who have all their genes in common) than those of unrelated people. Man has more scent glands than does any other primate, perhaps as a remnant of some uniqueness in smell which has lost its importance in a world full of sight. The tie between sex and scent in ourselves is made by a rare inborn disease that both prevents the growth of the sex organs and abolishes the sense of smell, suggesting that the two systems share a common pathway of development in the early embryo.

Variation in the way we look, see, smell and taste is but a tiny part of the universe of difference. The genes that enable mice to recognise each other by scent are part of a larger system of identifying outsiders. The threat of infection means that every creature is always in conflict with the external world. The immune system determines what should be kept out. It differentiates 'self' from 'not-self' and makes protective antibodies that interact with antigens (chemical clues on a native or foreign molecule) to define whether any substance is acceptable. The millions of antibodies each recognises a single antigen. Cells bear antigens of their own that, with great precision, separate each individual from his fellows. Antigens are a hint of the mass of uniqueness beneath the bland surface of the human race.

When blood from two people is mixed, it may turn into a sticky mess. The process is controlled by a system of antigens called the blood groups. Only certain combinations can mix successfully. Some groups, ABO and Rhesus for example, are familiar, while others, such as Duffy and Kell, are less so. Because of their importance in transfusion, millions of people have been tested. A dozen systems are screened on a routine basis and each comes in a number of forms. This small sample of genes generates plenty of diversity. The chances of two Englishmen having the same combination of all twelve blood groups is only about one in three thousand. Of an Englishman and a Welshman it is even less and of an English person and an African less again.

Since the discovery of the blood groups and other cues on the surfaces of cells, there has been a technical revolution. Like the stone age revolution a thousand centuries ago, it depends on simple tools that can be used in many ways. The DNA of different people can now be compared letter by letter, to test how unique we are. The Human Genome Diversity Project is a spin-off from the main mapping effort which has tested thousands of people. On the average, and depending on what piece of the DNA is tested, two people differ in about one or two DNA letters per thousand; that is, in about three to six million places in the whole inherited message. Some of the differences involve changes in single bases (single nucleotide polymorphisms, or 'snips' as they are called), some in the number of short repeats of particular sequences ('microsatellites' and 'minisatellites') and some turn on the presence or absence of bits of mobile DNA that leapt into a particular place in the genome long ago. Blood groups show how improbable it is that two will be the same when a mere twelve variable systems are used. The chance that they both have the same sequence of letters in the whole genetic alphabet is one in hundreds of billions. Genetics has made individuals of us all. It disproves Plato's myth of the absolute, that there exists one ideal form of human being, with rare flaws that lead to inborn disease.

Variation helps us to understand where we fit in our own family tree, in the pedigree of humankind, and in the world of life. Relatives are more likely to share genes because they have an ancestor in common. As all genes descend from a carrier long dead they can be used to test kinship, however distant that might be. The more variants two people share the more they are related. This logic can be used to sort out any pattern of affinity.

This detective work is easiest when close – or identical – relatives are involved. The US Army tests the fit of dead bodies to their previous owners by storing DNA samples from soldiers in the hope of identifying their corpses after death. DNA can also say a lot about the immediate family. Once, immigration officers faced with applicants for entry often refused to believe that a child was the offspring of the woman who claimed it. Comparison of the genes of mother and child almost always showed that the mother was telling the truth. Our society being what it is, the tests are now less used than they were. However, not all families are what they seem. Attempts to match the genes of parents and offspring in Britain or the United States reveal quite a high incidence of false paternity. Many children have a combination of genes which cannot be generated from those of their supposed parents. Often, they show that the biological father is not the male who is married to the biological mother. In middle class society about one birth in twenty is of this kind.

Such detective work can skip generations. During the Argentinian military dictatorship of the 1970s and 1980s thousands of people disappeared. Most were murdered. Some of the victims were pregnant women who were killed after they had given birth. Their children were stolen by military families. When civilian rule was restored, a group of mothers of the murdered women began to search for their grandchildren, whose DNA was compared with those who claimed to be their parents. The message passed in the genes enabled more than fifty children to be restored to their biological families, two generations on.

Other families have no hope of restoration. Bones dug up in a cellar in Ekaterinburg in 1991 were suspected to be those of the last Tsar and his family, shot in 1918. Checks of their DNA against modern relatives proves that the skeletons are, indeed, the remains of the Romanovs. Intriguingly enough, the skeleton of one young girl imprisoned with the group was missing. A woman known as Anna Anderson (who died in Virginia in 1984) claimed for many years to be the absent child, Anastasia, the daughter of the Tsar. Her assertion was rejected by a German court, but was accepted by thousands of émigré Russians. A check of the genes contained in a sample of her tissue found after her death showed her not to be related to the Romanovs, but instead to be (as many had suspected) a Pole, Franziska Schanzkowska, who had been rescued from a suicide attempt in a Berlin canal and ever after believed herself to be of noble blood.

Anna Anderson's claim to the Russian Eagle was false; but everyone has been granted a genetic coat of arms to democratize the search for descent. Like that of the Romanovs, it records who the forebears were and from whence they came. When people move they take more than their escutcheons. The DNA goes too, so that maps of genes do more than just record ancestry. They recreate history.

History itself may suggest where to start. Alex Haley, in his book *Roots*, used documents on the slave trade to try to find his African ancestors. He found just one, Kunta Kinte, who had been taken as a slave from the Gambia in 1767; and later became suspicious of the tales told to him by a native story-teller upon which *Roots* was in part based. The genes of today's Black Americans might have solved his problem.

The African slave trade began in the days of the Roman Empire. By AD 800 Arab traders had extended it to Europe, the Middle East and China. In the fifteenth century the Spanish and Portuguese started what became a mass migration, at first from the Guinea Coast, modern Mauritania. Mediaeval Venice had black gondoliers and by the sixteenth century one person in ten in Lisbon was of African origin. Soon, a bull of Pope Nicholas V instructed his followers to 'attack, subject, and reduce to perpetual slavery the Saracens, Pagans and other enemies of Christ, southward from Cape Bojador and including all the coast of Guinea'.

The main trade was to the New World. About fifteen million Africans were shipped across the Atlantic. They came from all over West Africa and were dispersed over much of North and South America. The United States imported less than a twentieth of the total, but by the 1950s the USA had a third of all New World people of African descent, suggesting that slaves were treated less brutally there than in the Caribbean or Brazil. Slave-owners had their own preferences. In South Carolina slaves from the Gambia were favoured over those from Biafra as the latter were thought to be hard to control. In Virginia the preference was in the opposite direction.

Many Africans have an abnormal form of the red pigment of the blood, haemoglobin. One of the amino acids has suffered a genetic accident, a mutation. This 'sickle-cell' form protects against malaria. Its protective role has disappeared with the control of the disease in the United States, but many thousands of Black Americans still carry the gene as an unwelcome record of their past. Anyone, however light their skin, who has the sickle-cell variant must have had at least one African ancestor. The disease was first recognised in 1910, and was at once used as a statement of racial identity: anyone with the illness (whatever their colour) must, by definition, be a Negro. Indeed, its very presence was seen as proof of the degenerate nature of American Blacks. The related disorders in southern Europe also showed, in the words of one racial theorist, that such people were 'not white clear through' and that their immigration to the USA would 'produce a hybrid race of people as worthless and futile as the good-for-nothing mongrels of Central America.'

The fact that many Black Americans have a copy of the gene for sickle-cell haemoglobin says little more than that they originated in West Africa, which we knew already. Molecular technology tells a tale of just who the 'mongrels' are. It uncovers a mass of variation around the haemoglobin genes and gives an insight into the ancestry of many Americans, black or not; including the great majority who do not carry a copy of sickle-cell at all.

The DNA in this part of the genome varies from place to place within Africa. The sickle-cell mutation itself is associated with different sets of DNA letters in Sierra Leone, Nigeria and Zaire, probably because it arose several times. The DNA around the normal version of the gene also varies and this, too, can be used to track down where in Africa the ancestors of today's Americans came from.

That continent contains more diversity than anywhere else. Not only are its people more distinct one from the other, but different villages, tribes and nations have more individuality, because humans have been in Africa for longer than anywhere else. As a result, genes can track down the ancestry of Africans with some accuracy.

Black Americans from the north of the USA have a different set of variants from those in the south. The majority of northerners share a heritage with today's Nigerians while their southern cousins have more affinities with peoples further west. The difference in the slave markets two hundred years ago has left evidence today. Alex Haley, by comparing his genes with those from Africa, would have learned much more about his forefathers than he could hope to uncover from the records. For any black American, a DNA test could be a first hint as to where to search for his slave ancestors – and, for a mere $250, one is now on sale (although the limited information yet available on the genes of West Africa mean that any hope of finding his native village – or even tribe – is largely vain).

Many of Alex Haley's ancestors were probably not black at all. One particular variant in the Duffy blood group system is found only in West Africa. Europeans have a different version of this gene. Surveys of United States Blacks show that up to a quarter of their Duffy genes are of white origin, in many cases because of inter-racial matings during the days of slavery. Such liaisons were covert, but widespread. Even President Thomas Jefferson is said to have had several children by his slave mistress, Sally Hemings. The conjecture was proved by the discovery that one of her descendants carries DNA shared with that of the President's family (a proof so firm that it has been accepted, grudgingly, by the association of Jeffersonian descendants).

A closer look at a set of DNA clues specific to Africa or to people of European origin says more about the history of slavery. In Jamaica (where whites were a small minority), just one black gene in sixteen is of European origin. In most American cities the figure is around one in six, but in New Orleans is higher, at between a fifth and a quarter. Until 1803, Louisiana was under French, rather than Anglo-Saxon, control. Gallic racial tolerance lives on in today's genes. The differences in numbers of blacks and whites, and the small proportion of white families that have mated with blacks has transferred far fewer black genes into the American population that sees itself as white, with an overall proportion of about one gene in a hundred.

Race involves a lot more than DNA. As a result, the proportion of blacks in the United States is rising. In 1997, about thirteen per cent of Americans perceived themselves as black and, over the past two decades the country's black population has increased at twice the rate of the white. Most of this has nothing to do with genes, but is a matter of identity. Thirty years ago anyone of mixed ancestry would do their best to classify themselves as white. Now, with the rise of black self-esteem, many find themselves more at home as blacks. As a result, any genetic measure of admixture then and now will give different results, as a reminder that race is constructed by society as much as by DNA.

Seventeenth- and eighteenth-century England, too, had a substantial black population. It disappeared; not because it died out, but because it was assimilated. Part of its heritage is, without doubt, still around in the streets of modern Britain. Dr Johnson himself had a black servant, Francis Barber, to whom he left enough money to set up in trade. Many people around Lichfield are proud to trace their descent from him, although their skins are as fair as those of their neighbours. White Britons contain other exotic genes as well. After all, the first slaves to cross the Atlantic were the Caribbean Indians sent to Spain by Columbus in 1495 and there was a sixteenth-century fashion for bringing newly discovered peoples back to Europe. The English explorer Frobisher brought back some Eskimos in 1577 and more than a thousand American Indians (including a Brazilian king) were transported to Europe. Many of the unwilling migrants died, but some brought up families. Their legacy persists, no doubt, today; but they have been absorbed so fully into the local population that only a genetic test – or provision of a dependable pedigree – can say who bears it.

Genes have taken us back for hundred of years – for fifteen generations or so where black Americans are concerned. But they bear messages from earlier in history. Sometimes, the evidence is direct, more often indirect: but in every case it links the present with the past.

For good historical reasons, a great deal is known about the genetics of Hiroshima and Nagasaki. The Americans spent many years on a survey of whether the atom bombs had increased the mutation rate. No effect was found, but a mass of information on the genes of the two cities was gathered. Each has a cluster of rare variants not present in the other. They are relics of an ancient history. Hiroshima and Nagasaki were each founded by the amalgamation of different warring clans that lived in the region eight thousand years ago. Like tribal peoples today, they had diverged in their DNAs. The slight differences between the ancient tribes persist in the modern towns. Nagasaki was one of the few ports open to the outside world during Japan's self-imposed isolation, but has no more sign of an influx of a foreign heritage than does Hiroshima. The voices of remote ancestors echo more loudly through the two cities than do those of more recent invaders.

Because genes copy themselves, there is no need to go back to the source to find an ancestor; but, sometimes, the source has been preserved. The Egyptian pharaoh Tutankhamun was buried at about the same time as another mummy, Smenkhare. Their blood groups can still be identified and show them to have been brothers. The first piece of human fossil DNA was found in the dried corpse of an Egyptian child, buried in the sands. It had survived for two and a half thousand years. Since then, many pieces of ancient DNA have turned up (although their analysis is confused by a tendency for contamination with modern material).

It has, nevertheless, become possible to read ancestral genes directly. Some ancient DNA, like that of the Easter Islanders, whose civilization was destroyed by constant warfare and ecological vandalism, has no equivalent in the modern world and remains, like their enigmatic statues, as the sole evidence of a people who left no posterity. Sometimes, it adds to the clues of the present. Agriculture began in Japan with the Jomon people, about ten thousand years ago, but they also spent much of their time as hunters. Farming did not take off as a way of life, with rice as a staple diet, until the Yayoi tribes who followed them, thousands of years later. Rice was brought by the Chinese, and the Japanese argue about how many of their genes entered the country with the crop. Many believe that the immigrants drove out most of the natives; that people moved, rather than ideas. However, DNA extracted from a two-thousand-year-old Chinese burial site links its inhabitants with modern Chinese, but not with the fossil DNA of the extinct Japanese. It proves that few mainlanders made the journey. Instead, the locals of two millennia ago, much like their modern descendants, picked up and used a new technology invented in a foreign land. Modern Japan, on the other hand, does have biological links with the Chinese, so that a movement from the mainland had an impact much later.

Some ancestral voices are particularly fluent in telling the story of the past. Mitochondria are small energy-producing structures in the cell. Each has its own piece of DNA, a closed circle of about sixteen thousand DNA bases, quite

distinct from that in the cell nucleus. Eggs are full of mitochondria but those in sperm are killed off as they enter the
egg. As a result, such genes are inherited almost exclusively
through females. Like Jewishness, they pass from mothers
to daughters and sons, but daughters alone pass them on
to the next generation.

Every family, every nation and every continent can trace
descent from its mitochondrial Eve, a woman (needless to
say, one of many alive at the same time) upon whom all
their female lineages converge. Sometimes she lived not
long ago: in New Zealand, for instance, nearly all Maoris
share the same mitochondrial identity, hinting that just a
few women founded their nation a thousand years ago. A
world family tree based on mitochondria finds its roots in
Africa, with more diversity in that continent than anywhere else. To track more recent paths of migration shows
that mitochondria are an accurate record of history: thus,
in the New World, native mitochondria have a tie with
those of Siberia, confirming an ancient pattern of
migration.

Shared genes link New Zealand, Siberia and the rest of
the world to an African ancestor. The first modern human
appeared in Africa over a hundred thousand years ago, in
the continent that gave rise to most of our pre-human kin
and of the apes to whom we claim affinity. A few of these
African relatives from a deeper branch of the tree are alive
today. One, the chimpanzee, has always seemed a near
neighbour; and Koko (an inhabitant of the Gombe Stream
Reserve) was the first animal to have an obituary in *The
Times*.

As any literate teenager knows, Tarzan of the Apes was
proved to be the son of Lord Greystoke by virtue of the
inky fingermarks in a childhood notebook. Galton had
shown that chimpanzees have fingerprints that look much
like those of a human being. Chimps and men, they prove,

share genes. A joint heritage goes beyond the fingertips. A distinguished geneticist of the 1940s once tested whether chimps share our variation in the ability to taste the bitter chemical PROP by feeding it to three of the inhabitants of London Zoo. Two swallowed the drink with every sign of delight, but the third spat the liquid all over the famous professor as further evidence of common ancestry.

The biological affinity goes much further. Apes have blood groups like our own, their chromosomes are almost identical, and a test of the overall similarity of DNA shows that humans share ninety-eight per cent of their genetic material with chimpanzees. We trace relatedness to the rest of the animal kingdom as well, with about a quarter of our genes similar to others in remote places among the insects or the jellyfish. Mice and men have much more in common, including dozens of inherited diseases. We share even more genes with rabbits and plenty with remote branches of existence, from bacteria to yeasts to bananas. All living creatures seem to need a set of 'housekeeping genes' that do the basic work of the cell, and many of the seven hundred such structures are shared. Most have changed little since they began. An unkind experiment in which more and more of the five hundred genes in a simple bacterium were destroyed showed that it needs, at an absolute minimum, three hundred or so; nearly all of which have parallels in our own DNA. This common core shows that the most unlikely beings speak the same genetic language.

Pharaoh Psamtik the First, who flourished in the seventh century before Christ, searched for the first word of all. He put a baby in the care of a dumb nurse and noted the sounds it made. One word was (or seemed to be) 'becos', the Phrygian for bread, suggesting to Psamtik that the Phrygians (who lived in what is modern Turkey) were the first people of all. A computer search through the millions of DNA letters now sequenced from dozens of organisms also hints at a shared structure from bacteria to humans; the father (or mother) of all genes, that might have persisted since life began. The scientist who published the ur-sequence has turned the information to a useful end. Assigning musical notes to each DNA letter he used them as a theme for a 'symphony of life'.

Gene sharing, from bacteria to humans, proves the unity of existence. It also defines the limits of what biology can say. A chimp may share ninety-eight per cent of its DNA with ourselves but it is not ninety-eight per cent human: it is not human at all – it is a chimp. And does the fact that we have genes in common with a mouse, or a banana, say anything about human nature? Some claim that genes will tell us what we really are. The idea is absurd.

One gene is found in a certain form in men, but a different one in all other apes. It codes for a molecule on the cell surface much involved in communication between cells, brain cells more than most. Perhaps this is the gene – or one of the genes – that makes us human. Its message spelt out in the four DNA letters, A, G, C and T starts like this: AACCGGCAGACAT . . . Altogether, it has three thousand letters. Together they contain an important part of the tedious biological story of being a man or woman rather than a chimpanzee or gorilla. Needless to say, that ancestral bulletin does nothing to tell us – or apes – what it means to be part of humankind. That calls for a lot more than a sequence of DNA bases and lies outside the realm of science altogether.

St Bede – whose writings are the best source of information about England before the eighth century – had a powerful metaphor for existence. To him human existence was 'As if when on a winter's night you sit feasting with your ealdormen and thegns, a single sparrow should fly swiftly into the hall, and coming in at one door instantly fly out through another. In that time in which it is indoors it is indeed not touched by the fury of the winter, but yet, this smallest space of calmness being passed almost in a flash, from winter going into winter again, it is lost to your eyes. Somewhat like this appears the life of man; but of what follows or what went before, we are utterly ignorant.'

His allegory was a religious one but has a biological parallel. Genes have a memory of their own. To read it gives new hope of looking beyond the hall into which our own brief existence is confined. It allows us to learn what went before in the life of our own species; to guess at what happened much earlier, and even to speculate about what fate may hold for generations yet to come.

Chapter Two

THE RULES OF THE GAME

It is always painful to watch an unfamiliar game and to try to work out what is going on. Although I lived in the United States for several years, and although the sport is now shown on British television, I have almost no idea how American football works. There is a clear general desire to score, but how play stops and starts and why the spectators cheer at odd moments remains a closed book. A deep lack of interest in ball games helps in my case, but cricket is equally dull to sporting enthusiasts from other countries. They just do not understand the rules.

The rules of the game known as sexual reproduction are not obvious from its results. As a consequence, how inheritance works was a closed book until quite recently. Part of the problem is that the way sex works is so different from how it seems that it ought to. It seems obvious that a character acquired by a parent must be passed on to the next generation. After all, blacksmiths' children tend to be muscular and those of criminals less than honest. In the Bible, Jacob, when allowed to choose striped kids from Laban's herd of goats, put striped sticks near the parents as they mated in the hope of increasing the number available. Later, pregnant women looked on pictures of saints and avoided people with deformities. It took a series of painful trials in which generations of mice were deprived of their tails to show that acquired characters were not in fact inherited. Of course, Jews had been doing the same experiment for thousands of years.

Another potent myth about inheritance is that the characters of a mother and a father pass to their blood, which is mixed in their offspring. Children are, as a result, a blend of the attributes of their parents. This idea – a sort of genetics of the average – copes quite well with traits such as height or weight but fails to explain why a child may look like a distant relative rather than its father or mother. The idea lasted until just a few years ago. The stud book is the record kept by racehorse breeders. A mare who had borne a foal by mating with a non-stud stallion was struck off as her blood was deemed to be polluted. Indeed, a survey of elderly women in Bristol showed that half believed in the chance of a woman having a black baby if she had sex with a black man many years before. The crones of the west country, like the breeders of horses, had never managed to work out the instructions for the reproductive game.

The only section of *The Origin of Species* which does not make good reading today is Chapter Five, 'Laws of Variation'. Darwin got it wrong and, after much agonising, suggested that the organs of parents passed material to the blood and then to sperm and egg. Children were, he thought, intermediate between those who produced them. Such a mode of inheritance would be fatal to the idea of evolution. The problem was pointed out by Fleeming Jenkin, the first Professor of Engineering at the University of Edinburgh. Writing in 1867 – and with a sturdy disregard of today's proprieties – Jenkin imagined 'a white man wrecked on an island inhabited by negroes. Suppose him to possess the physical strength, energy and ability of a dominant white race. There does not follow the conclusion that after a . . . number of generations the inhabitants of the island will be white. Our shipwrecked hero would probably become king; . . . he would have a great many wives, and children . . . much superior in average intelligence to the negroes, but can anyone believe that the whole island will gradually acquire a white or even a yellow population? A highly favoured white cannot blanch a nation of negroes.'

Jenkin saw that the attributes of a distant ancestor, valuable as they might be, are of little help to later generations if bloods mix. Characters would then blend over the years until their effects disappear. However useful an ink drop in a gallon of water might be at some time in the future it is impossible to get it back from a single mixed drop. Genetics by blending means that any advantageous character would be diluted out in the next generation. Fortunately, the blood myth is wrong.

It was shot down by Galton himself. He transfused blood from a black rabbit to a white to see if the latter had black offspring. It did not. Inheritance by dilution had been disproved, but Galton had nothing to put in its place.

Unknown to either Darwin or to his cousin the rules of genetics had already been worked out by another biological genius. Gregor Mendel lived in Bohemia and published in a rather obscure scientific journal, the *Transactions of the Brunn Natural History Society*. His breakthrough was overlooked for thirty-five years after it was published in 1866. Mendel, an Augustinian monk, attempted a science degree but failed to complete it. Like Darwin and Galton he suffered from bouts of depression which prevented him from working for months at a time. Nevertheless, he persisted with his experiments. He found that the inherited message is transmitted according to a simple set of regulations – the grammar of the genes. Later in his career (and setting a precedent for the present age) he was unable to continue with research because of the pressures of administration. The study of inheritance came to a halt for almost half a century.

Grammar is always more tedious than vocabulary, but

cannot be avoided. The rest of this chapter explores the basic rules of genetics. Those who teach the subject still have an obsession with Mendel and his peas and I make no excuse for having them as a first course.

Mendel made a conceptual breakthrough. Instead of (like his predecessors) working on traits such as height or weight (which could only be measured) Mendel was more or less the first biologist to count anything. This put him on the road to his great discovery.

Peas, like many garden plants, exist in true-breeding lines within which all individuals look the same. Different lines are distinct in characters such as seed shape (which can be round or wrinkled) and seed colour, which may be yellow or green. Peas also have the advantage that each plant carries both male and female organs. Using a small brush it is possible to fertilise any female flower with pollen from any male. Even a male flower from the same plant can be used. The process, a kind of botanical incest, is called self-fertilisation.

Mendel added pollen (male germ cells) from a line with yellow peas to the female part of a flower from a green pea line. In the next generation he got an unexpected result. Instead of all the offspring being intermediate, all the plants in the new generation looked like one of the parents and not the other. They all had yellow peas. This is not at all what would be expected if the 'blood' of the two lines was blended into a yellowish-green mixture.

The next step was to self-fertilise these first-generation yellow plants; in other words to expose their eggs to pollen from the same individual. That gave another unforeseen outcome. Both the original colours, yellow and green, reappeared in the next generation. Whatever it was that produced green could still do so, even though it had spent time within a plant with yellow peas. This did not fit at all with the idea that the different properties of each parent were blended together. Inheritance was, his experiment showed, based on particles rather than fluids.

Mendel did more. He added up the numbers of yellow and green peas in each generation. In the first generation (the offspring of the crossed pure lines) all the plants had yellow peas. In the second, obtained by self-fertilising the yellow plants from the first generation, there were always, on the average, three yellows to one green. From this simple result, Mendel deduced the fundamental rule of genetics.

Pea colour was, he thought, controlled by pairs of factors (or genes, as they became known). Each adult plant had two factors for pea colour, but pollen or egg received only one. On fertilisation – when pollen met egg – a new plant with two factors (or genes) was reborn. The colour of the peas was determined by what the plant inherited. In the original pure lines all individuals carried either two 'yellow' or two 'green' versions of the seed colour gene. As a result, crosses within a pure line gave a new family of plants identical to their parents.

When pollen from one pure line fertilised eggs from a different line new plants were produced with two different factors, one from each parent. In Mendel's experiment these plants looked yellow although each carried a hidden set of instructions for making green peas. In other words, the effects of the yellow version were concealing those of the green. The factor for yellow is, we say, *dominant* to that for green, which is *recessive*.

Plants with both variants make two kinds of pollen or egg. Half carry the instructions for making green peas and half for yellow. There are hence four ways in which pollen and egg can be brought together when two plants of this kind are mated, or a single one self-fertilised. One quarter of fertilisations involve yellow with yellow, one quarter green with green; and two quarters – one half – yellow with green.

Mendel had already shown that yellow with green

produces an individual with yellow peas. Yellow with yellow, needless to say, produces plants with yellow peas, and in a plant with two green factors the pea is green. The ratio of colours in this second generation is therefore three yellow to one green. Mendel worked backwards from this ratio to define his basic rule of inheritance.

Mendel made crosses using many different characters – flower colour, plant height and pea shape – and found that the same ratios applied to each. He also tested the inheritance of pairs of characters considered together. For example, plants with yellow and smooth peas were crossed with others with green and wrinkled peas. His law applied again. Patterns of inheritance of colour were not influenced by those for shape. From this he deduced that separate genes (rather than alternative forms of the same one) must be involved for each attribute. Both for distinct forms of the same trait (yellow or green colour, for example) and for quite different ones (such as colour and shape) inheritance was based on the segregation of physical units. Mendel was the first to prove that offspring are not the average of their parents and that genetics is based on differences rather than similarities.

Biologists since his day have delighted in picking over his results (and accusing him of fraud because they may fit his theories too well). They argue about what he thought his factors were, and speculate about why his work was ignored. Whatever lies behind its long obscurity, Mendel's result was rediscovered by plant breeders in the first year of the twentieth century and was soon found to apply to hundreds of characters in both animals and plants. Mendel had the good luck, or the genius, needed to be right where all his predecessors had been wrong. No science traces its origin to a single individual more directly than does genetics, and Mendel's work is still the foundation of the whole enormous subject which it has become.

Mendel rescued Darwin from his dilemma. A gene for green pea colour or for white skin, rare though it may be, is not diluted by the presence of many copies of genes for other colours. Instead, it can persist unchanged over the generations and will become more common should it gain an advantage.

Soon after the crucial rules were rediscovered they were used to interpret patterns of human inheritance. It is not possible to carry out breeding experiments on our fellow citizens. They would take too long, for one thing. Instead, biologists must rely on the experiments which are done as humans go about their sexual business. They use family trees or pedigrees – from the French *pied de grue*, crane's foot, after a supposed resemblance of the earliest aristocratic pedigrees (which were arranged in concentric circles) to a bird's toes. Some are fanciful, going back to Adam himself, but geneticists usually have fewer generations to play with, although one or two pedigrees do trace back for hundreds of years.

The first was published in 1903. It showed the inheritance of shortened hands and fingers in a Norwegian village. Such fingers ran in families and showed a clear pattern. The trait never skipped a generation. Anyone with short fingers had a parent, a grandparent and so on with the same thing. If an affected person married someone without the abnormality (as most did), about half their children were affected. If any of their normal children married another person with normal hands the character disappeared from that branch of the family.

The pattern is just what we expect for a dominant character. Only one copy of the damaged DNA (as in the case of yellow pea colour) is needed to show its effects. Most sufferers, coming as they do from a marriage between a normal and an affected parent, have a single copy of the normal and a single copy of the abnormal form, one from either parent. As a result, their own sperm – or eggs – are of two types, half carrying the normal and half the abnormal variant. When they marry, half their children carry a copy of the damaged gene. The chance of any child of a normal and an affected person having short fingers is hence one in two. An unaffected couple never has a child showing the abnormality as neither of them possesses the flawed instruction that makes it.

Other inherited traits do not behave in this simple way. They are recessives. To show the effect, two copies of the inherited factor, one from each parent, are needed. The parents themselves usually each have a single copy and appear quite normal. Most do not know that they are at risk of having an affected child. Sometimes, though, their offspring looks more like a distant relative or an ancestor than it does either parent. Before Mendel, that pattern was inexplicable. Such children were sometimes called 'throwbacks'. Now we know that they are obeying Mendel's laws. They have, by chance, inherited two copies of a recessive abnormality while their mother and father each have just one.

In Britain, one child in several thousand is an albino, lacking any pigment in eyes, hair or skin. Elsewhere, the anomaly is more common. In some North American Indians, about one person in a hundred and fifty is an albino. According to the Book of Enoch (one of the apocryphal books of the Bible), Noah himself suffered from the condition. If he did, there is not much sign of the gene in his descendants.

The great majority of albino children are born to parents of normal skin colour. They must each have a single copy of the albino factor matched with another copy of that for full pigmentation. Half the father's sperm carry the altered gene. Should one of these fertilise one of that half of his partner's eggs which carry the same thing, then the child will have two copies of the recessive form and will lack pigment. In a marriage such as this, the chance of any child being an albino is a half times a half. This one in four probability is the same for all the children. It is not the case, as some parents think, that having had one albino child means that the next three are bound to be normal.

Patterns of inheritance in humans can, then, follow the same rules as those found in peas. However, biology is rarely pure and never simple. Much of the history of human genetics has been a tale of exceptions to Mendel's laws.

For example, variants do not have to be dominant or recessive. In some blood groups, both show their effects. Someone with a factor for group A and group B has AB blood, which shares the properties of both. At the DNA level, the whole concept of dominance or recessivity goes away. A change in the order of bases can be identified with no difficulty, whether one or two copies are present. Molecular biology makes it possible to see genes directly, rather than having to infer what is going on, as Mendel did, from looking at what they make.

Another result which would have surprised Mendel is that one gene may control many characters. Thus, sickle-cell haemoglobin has all kinds of side-effects. People with two copies may suffer from brain damage, heart failure and skeletal abnormalities (all of which arise from anaemia and from the blockage of blood vessels). In contrast, some characters (such as height or weight) are controlled by many genes. What is more, Mendelian ratios sometimes change because one or other type is lethal, or bears some advantage.

All this (and much more) means that the study of inheritance has become more complicated in the past century and a half. Nevertheless, Mendel's laws apply to humans as much as to any other creature.

They are beguilingly simple and have been invoked to explain all conceivable – and some inconceivable – patterns of resemblance. In the early days, long pedigrees claimed to show that outbursts of bad temper were due to a dominant gene and that there were genes for going to sea or for 'drapetomania' – pathological running away among slaves. This urge for simple explanations persists today, but mainly among non-scientists. Geneticists have had their fingers burned by simplicity too often to believe that Mendelism explains everything.

Mendel had no interest in what his inherited particles were made of or where they might be found. Others began to wonder what they were. In 1909 the American geneticist Thomas Hunt Morgan, looking for a candidate for breeding experiments hit upon the fruit fly. It was an inspired choice and his work, with *Drosophila melanogaster* (the black-bellied dew lover, to translate its name) was the first step towards making the human gene map.

Many fruit fly traits were inherited in a simple Mendelian way, but some showed odd patterns of inheritance. When peas were crossed it made no difference which parent carried green or yellow seeds. The results were the same whether the male was green and the female yellow, or vice versa. Some traits in flies gave a different result. For certain genes – such as that controlling the colour of the eye, which may be red or white – it mattered whether the mother or the father had white eyes. When white-eyed fathers were crossed with red-eyed mothers all the offspring had red eyes but when the cross was the other way round (with white-eyed mothers and red-eyed fathers) the result was different. All the sons had white eyes and the daughters red. To Morgan's surprise, the sex of the parent that bore a certain variant had an effect on the appearance of the offspring.

Morgan knew that male and female fruit flies differ in

another way. Chromosomes are paired bodies in the cell which appear as dark strands. Most of the chromosomes of the two sexes look similar but one pair – the sex chromosomes – are different. Females have two large X chromosomes; males a single X and a much smaller Y.

Morgan noticed that the pattern of inheritance of eye colour followed that of the X chromosome. Males, with just a single copy of the X (which comes from their mother, the father providing the Y) always looked like their mother. In females, the copy of the X chromosome from the mother was accompanied by a matching X from the father. In a cross between white-eyed mothers and red-eyed fathers, the female offspring have one X chromosome bearing 'white' and another bearing 'red'. Just as Mendel would have expected, they have eyes like only one of the parents, in this case the one with red eyes.

The eye colour gene and the X chromosome hence show the same pattern of inheritance. Morgan suggested that this meant that the gene for eye colour was actually on the X chromosome. He called this pattern 'sex-linkage'. Chromosomes were already candidates as the bearers of genes as, like Mendel's hypothetical particles, their number is halved in sperm and egg compared to body cells.

Everyone has forty-six chromosomes in each body cell. Twenty-two of these are paired, but the sex chromosomes, X and Y, are distinct. Because the Y carries few genes, in males the ordinary rules of Mendelian dominance and recessivity do not apply. Any gene on the single X will show its effects in a male, whether or not it is recessive in females.

The inheritance of human colour blindness is just like that of *Drosophila* eye colour. When a colour-blind man marries a normal woman none of his children is affected, but a colour-blind woman whose husband has normal vision passes on the condition to all her sons but none of her daughters. Because all males with the abnormal X show its effects (while in most females the gene is hidden by one for normal vision) the trait is commoner in boys than in girls. Many other abnormalities show the same pattern.

Sex-linkage leads to interesting differences between the sexes. For the X chromosome, females carry two copies of each gene, but males only one. As a result, women contain more genetic information than do men. Because of the two different sensors for the perception of red controlled by a gene on the X chromosome, many women must carry both red receptors, each sensitive to a slightly different point in the spectrum. Males are limited to just one. As a result, some women have a wider range of sensual experience – for colour at least – than is available to any man.

Whatever the merits of seeing the world in a different way, women have a potential problem with sex-linkage. Any excess of a chromosome as large as the X is normally fatal. How do females cope with two, when just one contains all the information needed to make a normal human being (or a male)? The answer is unexpected. In almost every cell in a woman's body one or other of her two X chromosomes is switched off.

Tortoiseshell cats have a mottled appearance, which comes from small groups of yellow and black hairs mixed together. All tortoiseshells are females and are the offspring of a cross in which one parent passes on a gene for black and the other transmits one for yellow hair. Because the coat-colour gene is sex-linked about half the skin cells of the kitten switch off the X carrying the black variant and the remainder that for yellow. The coat is a mix of the two types of hair, the size of the patches varying from cat to cat.

The same happens in humans. If a woman has a colour-blind son, she must herself have one normal and one abnormal colour receptor. When a tiny beam of red or green light is scanned across her retina her ability to tell the colour of the light changes as it passes from one group of cells to the next. About half the time, she makes a perfect match but for the rest she is no better at telling red and green apart than is her colour-blind son. Different X chromosomes have been switched off in each colour-sensitive cell, either the normal one or that bearing the instruction for colour blindness.

The inheritance of mitochondrial genes also shows sexual differences. When an egg is fertilised, much of its contents, including those crucial structures, is passed on to the developing embryo. Mitochondria have a pattern of inheritance quite different from those in the nucleus. They do not bother with sex, but instead are passed down the female line. Sperm are busy little things, with a long journey to make, and are powered by many mitochondria. On fertilisation these are degraded, so that only the mother's genes are passed on. In the body, too, mitochondria are transmitted quite passively, each cell dividing its population among its descendants. Their DNA contains the history of the world's women, with almost no male interference. Queen Elizabeth the Second's mitochondrial DNA descends, not from Queen Victoria (her ancestor through the male line) but from Victoria's less eminent contemporary Anne Caroline, who died in 1881.

Mitochondria, small as they are, are the site of an impressive variety of diseases. Their sixteen and a half thousand DNA bases – less than a hundredth of the whole sequence – were, a century after the death of Anne Caroline, the first to be read off. Every cell contains a thousand or so of the structures. They are the great factories of metabolism; places where food – the fuel of life – is burned. Mitochondrial genes code for just thirteen proteins, and about twice that number of the molecules that transfer information from the DNA to where proteins are made.

They are more liable to error than are others. Some of the mistakes pass between generations, while others build up in the body itself as it ages. Some of the two hundred known faults involve single changes in the DNA, others the destruction of whole lengths of genetic material. Some are frequent: thus, a certain change in one mitochondrial gene is present in about one in seven thousand births.

Mitochondrial disease involves many symptoms: deafness, blindness, or damage to muscles or the brain. Certain forms of diabetes are due to mitochondrial errors, as is an inherited muscle weakness and drooping of the eyelids. Different patients in the same family may have distinct problems; perhaps deafness in one child and brain damage in another. All this comes from the role of mitochondria in burning energy and from their random shuffling as cells divide. An egg may carry both normal and abnormal mitochondria. If, in an embryo, those with an error become by chance common in the cell lines that make brain tissue, that organ suffers; if in cells that code for insulin, then diabetes is the result. Mothers pass such genes to sons and daughters, but only daughters pass it to the next generation; a pattern quite different from sex-linked inheritance.

These, then, are the rules of the genetical game. From here on, the rest is molecular biology: mechanics rather than physics. The notion that life is chemistry came first from humans. In 1902, just two years after the rediscovery of Mendelism, the English physician Sir Archibald Garrod noticed that a disease called alkaptonuria – at the time thought to be due to an intestinal worm – was more frequent in the children of parents who shared a recent ancestor than in those of unrelated people. Its symptoms, a darkening of the urine and the earwax, together with arthritis, followed that of a recessive. The disease was, he thought, due to an inherited failure in one of the pathways of metabolism, what he called a "chemical sport' (Darwin's own word for a deviation from the norm). It was the first of many inborn errors of metabolism. The actual gene itself was found just four years before the century ended. The key to its discovery showed how wide the genetical net must spread. An identical was found in a fungus, and that piece of damaged DNA used to search out its human equivalent.

What genes are made of came from the discovery it was possible to change the shape of bacterial colonies by inserting a 'transforming principle' extracted from a relative with different shaped colonies. That substance was DNA, discovered many years before in some rather disgusting experiments using pus-soaked bandages. It was the most important molecule in biology.

The story of how the structure of DNA, the double helix, was established is too well known to need repeating. The molecule consists of two intertwined strands, each made up of a chain of chemical bases – adenine, guanine, cytosine and thymine – together with sugars and other material. The bases pair with each other, adenine with thymine and guanine with cytosine. Each strand is a complement of the other. When they separate, one acts as the template to make a matching strand. The order of the bases along the DNA contains the information needed to produce proteins. Every protein is made up of a series of different blocks, the amino acids. The instructions to make each amino acid are encoded in a three-letter sequence of the DNA alphabet.

The inherited message contained within the DNA is passed to the cytoplasm of the cell (which is where proteins are made) through an intermediary, RNA. This ribose-nucleic acid comes in several distinct forms, each involved in passing genetic information to where it is used.

The DNA molecule – the agent of continuity between generations – has become part of our cultural inheritance.

The new ability to read (and to interfere with) its message has transformed our vision of our place in nature and our dominion over its inhabitants. It is, nevertheless, worth remembering that the laws of genetics were worked out with no knowledge of where or what the inherited units might be. Like Newton, Mendel had no interest in the details. He was happy with a universe of interacting and independent particles which behaved according to simple rules. These rules worked well for him, and often work just as well today.

Again like Newton, Mendel was triumphantly right, but only up to a point. Molecular biology has turned a beautiful story based on peas into a much murkier tale which looks more like pea soup. The new genetical fog is described in the next chapter.

Chapter Three

HERODOTUS REVISED

The Greek traveller Herodotus felt that he knew the world well. He voyaged around the Mediterranean and heard much of the Phoenicians' journeys into Africa. By putting what he knew of the globe's landmarks together he came to the conclusion that 'Europe is as long as Africa and Asia put together, and for breadth is not, in my opinion, even to be compared with them.' Herodotus had things in about the right places in relation to each other but the physical distances between them were hopelessly wrong.

For two thousand years maps could only be made in the Greek way. They were relative things, made by trying to fit landmarks together, with no measure of the absolute distances involved. Familiar bits of the countryside loomed far larger than they deserved. Mediaeval charts were not much better. Although the shape of Africa is recognisable it is much distorted. The cartographers' perception of remoteness was determined by how long it took to travel between two points rather than how far apart they really were.

Genetics, like geography, is about maps; in this case the inherited map of ourselves. Not until the invention of accurate clocks and compasses two thousand years after Herodotus was it possible to measure real distances on the earth's surface. Once these had been perfected, good maps soon appeared and Herodotus was made to look somewhat foolish. Now the same thing is happening in biology. Geneticists, it appears, were until not long ago making the same mistakes as the ancient Greeks.

Just as in mapping the world, progress in charting genes had to wait for technology. Now that it has arrived the shape of the biological atlas has been revolutionised, with a change in world-view far greater than that which separates the geography of the Athenians from that of today. What, even three decades ago, seemed a simple and reliable chart of the genome (based, as it was, on landmarks such as the colour of peas or of inborn disease) now looks very deformed.

The great age of cartography was driven, in the end, by economics: by the desire to find new materials and new markets. The mappers' Columbian ambitions needed a Ferdinand and Isabella. Even fifty years ago, to those in the know, there seemed to be money in DNA, and many great foundations gave cash to the subject. Not until the 1980s did it seem feasible to chart the whole lot and, even then, it seemed that the task would take decades. Such is the rate of progress that the job is now, just after the millennium, in effect complete. The politician's ear and the scientist's ego shifted cash into Programs, Institutes and Centres as the free market in science was abandoned in favour of the planned economy; but, in the end, the Human Genome Project worked and at last we have the map of ourselves. Taxpayers (most of them American) played an important part, but in its latter days the job was split, with some acrimony, between governments in consort with charities (such as the Wellcome Foundation at its campus near Cambridge) and private institutions, the biggest run by a defector from an American government laboratory. There was a mad rush to patent genes. Large sums changed hands. The rights to one technology were sold to a Swiss company for three hundred million dollars. At the end of the DNA bonanza the altruists were ahead and large parts of the information were fed onto the internet, where it is available to all.

The idea of a gene map came first not from technology but from deviations from Mendel's laws. Morgan, with his flies, found lots of inherited attributes that followed the rules. Their lines of transmission down the generations were not connected to each other; like pea colour and shape the traits were independently inherited. There was one big exception. Certain combinations of characters, those on the sex chromosomes, did not behave in this way. Soon, they were joined by others.

Mendel found that the inherited ratios for the colour of peas were not affected by whether the peas were round or wrinkled. Morgan, in contrast, discovered that, quite often, pairs of characteristics (such as eye colour and sex) travelled down the generations together. Soon, many different genes (such as those for eye colour, reduced wings and forked body hairs) in flies were found to share a pattern of inheritance with sex and, as a result, with the X chromosome. They were, in flagrant disregard of Mendel's rules, not independent. To use Morgan's term, they were linked.

Within a few years, many other traits turned out to be transmitted together. Experiments with millions of flies showed that all *Drosophila* genes could be arranged into groups on the basis of whether or not their patterns of inheritance were independent. Some combinations behaved as Mendel expected. For others, pairs of traits from one parent tended to stay together in later generations. The genes involved were, as Morgan put it, in the same linkage group. The number of groups was the same as the number of chromosomes. This discovery began the 'linkage map' of *Drosophila* and became the connection between Mendelism and molecular biology.

Linkage is the tendency of groups of genes to travel together down the generations. It is not absolute. Genes may be closely associated or may show only a feeble preference for each other's company. Such incompleteness is explained by some odd events when sperm and egg are formed. Every cell contains two copies of each of the chromosomes. The number is halved during a special kind of cell division in testis or ovary. The chromosomes lie together in their pairs and exchange parts of their structure. Sperm or egg cells hence contain combinations of chromosomal material that differ from those in the cells of the parents who made them.

That is why, within a linkage group, certain genes are inherited in close consort while others have a less intimate association. If genes are near each other they are less likely to be parted when chromosomes exchange material. If they are a long way apart, they split more often. Pairs of genes that each follow Mendel are on different chromosomes. Recombination, as the process is called, is like shuffling a red and a black hand of cards together. Two red cards a long way apart in the hand are more likely to find themselves split from each other when the new deck is divided than are two such cards close together. Such rearrangements mean that each chromosome in the next generation is a new mixture of the genetic material made up of re-ordered pieces of the chromosome pairs of each parent.

Recombination helped make the first genetic maps. Like the cards in a hand held by a skilled player, genes are arranged in a sequence. Their original position can be determined by how much this is disturbed each generation as the inherited cards are shuffled. By studying the inheritance of groups of genes Morgan worked out their order and their relative distance apart. Combining the information from small sets of inherited characters allowed what he called a 'linkage map' to be made.

Linkage maps, based as they are on exceptions to Mendelism, are very useful. They have been made for bacteria, tomatoes, mice and many other beings. Thousands of genes have been mapped in this way. In *Drosophila* almost all have been arranged in order along the chromosomes and in mice almost as many.

Because this work needs breeding experiments, the human linkage map remained for many years almost a perfect and absolute blank. Most families are too small to look for deviations from Mendel's rules and too few variants were known to look for them. There seemed little hope that a genetic chart of humankind could be made.

The one exception to this *terra incognita* was sex linkage. If genes are linked to the X chromosome, they must be linked to each other. It did not take long for dozens of traits to be mapped there. To draw the linkage map for other chromosomes was a painfully slow business. The gene for colour-blindness was mapped to the X in 1911, but the first linkage on other chromosomes did not emerge until 1955, when the gene for the ABO blood groups was found to be close to that for an abnormality of the skeleton. The actual number of human chromosomes was established in the following year and the first non-sex linked gene mapped onto a specific chromosome in 1968.

Now, genetics has been transformed. The technology involved is as to linkage mapping as satellites are to sextants. It does not depend on crosses and comes up with much more than a biological chart based on patterns of inheritance. Geneticists have now made a more conventional (but much more detailed) kind of chart, a physical map of the actual order of all the bases along the DNA. The new atlas of ourselves has changed our views of what genes are.

In the infancy of human genetics, thirty years ago, biologists had a childish view of what the world looks like. As in the mental map of an eleven year-old (or of Herodotus) linkage was based on a few familiar landmarks placed in relation with each other. The tedious but objective use of a measure of distance changed all that. Thirty years ago, molecular biologists were full of hubris. They had, they thought, solved the problems of inheritance. The new ability to read the DNA message would do the job that family studies and linkage mapping had failed to complete; it would show where all our genes were in relation to each other. The edifice whose foundations were laid by Mendel would then be complete. Optimism was, at the time, reasonable. It seemed a fair guess that the physical map of the genes would look much like a biological map based on patterns of inheritance and might in time replace it.

Such optimism was soon modified. The first explorations of the unknown territory which lay along the DNA chain showed that the physical map was quite different from the linkage map as inferred from peas or fruit-flies. The genes themselves are not beads lined up on a chromosomal string, but have a complicated and unexpected structure.

The successes of the molecular explorers depended, like those of their geographical predecessors, on new surveying instruments which made the world a bigger and more complicated place. The tools used in molecular geography deserve a mention.

The first device is *electrophoresis*, the separation of molecules in an electric field. Many biological substances, DNA included, carry an electrical charge. When placed between a positive and a negative terminal they move towards one or the other. A gel (which acts as a sieve) is used to improve the separation. Gels were once made of potato starch, while modern ones are made of chemical polymers. I have tried strawberry jelly, which works quite well. The gel separates molecules by size and shape. Large molecules move more slowly as they are pulled through the sieve while smaller ones pass with less difficulty. Various tricks improve the process. Thus, a reversal of the current every few seconds means that longer pieces of DNA can be electrophoresed, as they wind and unwind each time the power is interrupted. The latest technology uses arrays of fine glass tubes filled with gel, into each of which a sample is loaded. With various tricks the whole process becomes a production line and tens of thousands of samples can be analysed each day.

The computer on which I wrote this book has some fairly useless talents. It can – if asked – sort all sentences by length. This sentence, with its twenty words, would line up with many otherwise unrelated sentences from the rest of the book. Electrophoresis does this with molecules. The length of each DNA piece can be measured by how far it has moved into the gel. Its position is defined with ultra-violet light (absorbed by DNA), with chemical stains, fluorescent dyes that light up when a laser of the correct wavelength is shone on them, or with radioactive labels. Each piece lines up with all the others which contain the same number of DNA letters.

Another tool uses enzymes extracted from bacteria to divide the landscape into manageable pieces. Bacteria are attacked by viruses which insert themselves into their genetic message and force the host to copy the invader. They have a defence: enzymes which cut foreign DNA in specific places. These 'restriction enzymes' can be used to slice human genes into pieces. Dozens are available, each able to cut a particular group of DNA letters. The length of the pieces that emerge depends on how often the cutting-site is repeated. If each sentence in this volume was severed whenever the word 'and' appeared, there would be thousands of short fragments. If the enzyme recognised the word 'but', there would be fewer, longer sections; and an enzyme that sliced through the much less frequent word 'banana' (which, I assure you, does appear now and again) would produce just a few fragments thousands of letters long.

The positions of the cuts (like those of the words and, but and banana) provide a set of landmarks along the DNA. To track them down is a first step to reconstituting the book itself. The process is close to that carried out by the students who stormed the American Embassy in Tehran after the fall of the Shah. With extraordinary labour they pieced together secret documents which had been put through a shredder. By putting the fragments together the students reconstituted a long, complicated and compromising message.

Molecular biology does much the same. First, it needs to multiply the number of copies of the message to allow each short piece to be surveyed in detail as a preliminary to the complete map. Various tricks allow cut pieces of DNA to be inserted into that of a bacterium or yeast. The DNA has been cloned. Whenever the host divides, it multiplies not only its own genetic message but the foreign gene. As a result, millions of copies of an original are ready for study in the exquisite detail needed for genetic geography.

Cloning has been supplemented by another contrivance, the polymerase chain reaction. This takes advantage of an enzyme used in the natural replication of DNA to make replicas of the molecule in the laboratory. To pursue our rather tortured literary analogy, the method is a biological photocopier which can produce many duplicates of each page in the genetic manual. The photocopying enzyme comes from a bacterium which lives in hot springs. The reaction is started with a pair of short artificial DNA sequences which bind to the natural DNA on either side of the length to be amplified. By heating and cooling the reaction mixture and feeding it with a supply of the four bases, the targeted strands of DNA unwind, copy themselves with the help of the enzyme, and re-form. Each time the cycle is repeated, the number of copies doubles and millions of replicas of the original piece of DNA are soon generated.

Another piece of trickery exploits DNA's ability to bind to a mirror image of itself. DNA bases form two matched pairs; A with T and G with C. To find a gene, a complementary copy is made in the laboratory. When added to a cell this seeks out and binds to its equivalent on the chromosome. My computer can do the same. On a simple command, it will search for any word I choose and highlight it in an attractive purple. It does the job best with rare words (like 'banana'). A DNA probe labelled with a fluorescent dye shows up genes in the same way. The method is known as FISHing (for Fluorescent In-Situ Hybridisation) for genes. A modified kind of FISH involves unwinding the DNA before it is stained. This makes the method more sensitive.

All this and much more has revolutionised the mapping of human DNA. First, it has improved the linkage map. Patterns of inheritance of short sequences of DNA can be tracked through the generations just as well as can those of colour-blindness or stubby fingers. There are millions of sites which vary from person to person. All can be used in pedigree studies. Another scheme is to use the polymerase chain reaction to multiply copies of DNA from single sperm cells. The linkage map is made from a comparison of the reordered chromosomes in the sperm with that in the man who made them. This avoids the problem of family size altogether.

Linkage mapping in humans took a long time to get started and still has some way to go. Before the days of high technology the great problem was a shortage of differences; of variable genes, or segments of genetic material, whose joint patterns of inheritance could be studied. That problem has been solved. Our DNA is now known to be saturated with hundreds of thousands of variable sites, many based on individual variation in the numbers and positions of repeats of the two letters C and A. As a result, a whole new industry based on the most traditional kind of genetics has burst into existence.

It needs, like any industry, raw material. The French, together with the Americans, have identified sixty or so large families with long and complicated pedigrees, well suited for gene mapping. They come from various parts of the world, from Venezuela to Bangladesh. From each individual, lines of cells are kept alive in the laboratory and thousands of variants have been identified, tightly packed along the entire length of the chromosomes. Patients with, say, heart disease can be screened to see whether they also tend to carry other inherited variants. If they do, there is a good chance that the actual gene involved is nearby, and is dragging its anonymous fellows along with it. To find such a milestone may be the first step to the gene itself.

The descendants of Morgan have at last managed to do for humans what was long ago achieved for the fruit fly, and a linkage map of man is close at hand. That of woman, it transpires, is rather longer. Such maps depends on the sexual reshuffling of genes. This takes place, for some reason, more in females than in males and, as a result, their chart works to a different scale.

The human linkage map is useful, but biologists have always wanted to make a different kind of chart, one rather like that used by geographers, based on a straightforward description of the genetic material. Now, it is here. The approach was brutal: to assault the genome with time, money and tedium until the whole lot was read from one end to the other.

The first move in tying the linkage map to one based on the physical structure of DNA depended on a stroke of luck. Morgan noticed that in one of his fly stocks a gene which was usually sex-linked started behaving as if it was

not on the X chromosome at all. A glance down the microscope showed why. The X was stuck to one of the other
chromosomes and was inherited with it. A change in the
linkage relationships of the gene was due to a shift in its
physical position.

Such chromosomal accidents were used to begin the
human physical map. Sometimes, because of a mistake in
the formation of sperm or egg, part of a chromosome shifts
to a new home. Any parallel change in the pattern of
inheritance of a particular gene shows where it must be.
Now and again a tiny segment of chromosome is absent.
That can lead to several inborn diseases at once. One
unfortunate American boy had a deficiency of the immune
system, a form of inherited blindness, and muscular dystrophy. A minute section of his X chromosome had been
deleted. It must have included the length of DNA which
carried these genes. He gave a vital hint as to just where
the gene for muscular dystrophy – one of the most frequent
and most distressing of all inherited diseases – was located.
The absent segment was a landmark upon which a physical
map of the area around this gene could be anchored.

To map genes with changes in chromosomes need not
wait for natural accidents. Human cells, or those of mice
or hamsters, can be cultured in the laboratory. When mixtures of mouse and human cells are grown together, the
cells may fuse to give a hybrid with chromosomes from
both species. As the hybrids divide, they lose the chromosomes (and the genes) from one species or the other. Some
specifically human genes are lost each time a human
chromosome is ejected. To match the loss of particular
genes with that of chromosomes (or of their short segments) shows where they must be.

All these methods hint at a gene's position rather than
giving its precise coordinates. Small-scale cartography (or
mindless sequencing, as it is affectionately known) involves

various clever ruses. One depends on the ability of DNA to copy itself when a special enzyme is provided and the mixture fed with the A, G, C and T bases. It is possible to gradually lengthen pieces of a DNA strand from one end to the other, in four separate experiments (each using a different base). By chemical trickery, some of the growing strands are stopped each time a base is added. This produces a set of DNA pieces of different length, each stopped at an A, a G, a C or a T. Electrophoresis of the mixtures on the same gel gives four parallel lines of DNA fragments arranged by length. A scan across and down the gel gives the order of the bases. This is a most tedious task. It has been supplanted by machines that do the job in other ways. The most important change in genetics is a conceptual one. Because the three-letter code for each amino acid is known, it is possible to deduce the order of the amino acids made by a piece of the DNA once its sequence of bases has been established. What any gene does can be inferred by comparing that sequence with the computer database of others whose job is known. The fit need not be precise; after all, a French dictionary contains thousands of words similar enough to those in English to allow its meaning to be guessed at. It is also sometimes possible to work out the three-dimensional structure of the protein from its amino acid sequence and to deduce what its function might be.

There are some remarkable similarities among inherited vocabularies. The genes that control development are similar in humans and fruit flies, as are those that make their brains. Genes that, when they go wrong, damage the nervous system have close analogues in yeast (which do not have nerves at all) and one of our own genes is almost identical to another that alters the pattern of veins on an insect wing. Such conservatism has had a radical influence on human genetics.

The parts catalogue for a Mercedes C-class car contains four and a half thousand named items, from accelerator pedal to wing mirror to wheel nuts. Some (like individual bolts or washers) may be repeated dozens of times; but the factory has to make fewer than five thousand pieces to feed its assembly line and, in the end, to make its contribution to the European traffic jam. To make a human takes ten times as many – an executive jet's worth – and the task of seeing how that vast number of pieces is bolted together might seem almost impossible. Even the yeast cell (scarcely the Mercedes of the living world) needs more than the car, with six thousand proteins.

The yeast gene sequence itself, like any other, is no more than a factory manual, containing information on castings, mouldings and blanks but also on various extraneous bits which are removed before the assembly line gets them. Then, as in the Mercedes factory, the parts have to be put together to make a functional piece of machinery. Even that is of no use to someone who cannot drive, and even a skilled driver is no help when dumped in a strange city without a road map. To understand the workings of the cell demands even more.

DNA dismantlers, like car wreckers, generate only a box of bits and pieces; the biological equivalents of the nuts, bolts, relays, springs, struts, wires and all the other things needed to make an automobile. The shape of a human protein can be inferred from a DNA sequence, but even usually gives no hint as to how it fits into the cellular machinery. Yeast are simpler, and rather more is known about their mechanics. Life's unwillingness to change allows the yeast machine to be used to explore our own cells. One approach in the human gene hunt is rather like fishing. Take a protein whose job is known, and attach a molecular hook and a separate float to it. Insert it into a male (or a cell showing what passes for maleness in yeasts).

Then, mate that alluring individual to a female and drift his gene past all her thousands of cell parts until one takes the bait by slotting into it. The float causes the female cell to light up and the match is made.

A fishing expedition with two hundred or so bait proteins from yeast captured more than a thousand genes in human cells. One whole set of yeast proteins attached themselves to a single human protein that tells the cell when to start dividing and when to stop. The yeast bait is similar to one that, when it goes wrong, causes human cancer: and a quick test proved that the newly hooked human equivalents represented crucial parts of our own cells' brake and accelerator systems. Such a discovery is of great interest to medicine, and marked the first step in what may become an era of hunting for genes in complex creatures with a lure based on more humble beings.

The genetic languages spoken by different organisms are close indeed; close enough, in fact, to give an even chance that a newly-discovered human gene sequence will be related to something else, either another of our genes or one from a creature remote from ourselves. Human genetics has been transformed. No longer does it start with an inherited change (such as a genetic disease) and search for its location. Instead, it uses the opposite strategy, with a logic precisely opposite that of Mendel: from inherited particle to function, rather than the other way around. Genetics is the first science to have accelerated by going into reverse.

The first breakthrough of this new approach was the successful hunt for the cystic fibrosis gene in 1990. It gave a hint as to what was possible and was the introduction to the advances that led to the complete map a mere decade or so later. The job cost one hundred and fifty million dollars, but the costs per gene have now dropped by hundreds of times.

Cystic fibrosis is the most common inherited abnormality among white-skinned people. In Europe, it affects about one child in two thousand five hundred. Until a few years ago those with the disease died young. Their lungs filled with mucus and became infected. Those with the illness find it hard to digest food as they cannot produce enough gut enzymes. Its dangers have long been recognised. Swiss children sing a song that says 'The child will die whose brow tastes salty when kissed.' These symptoms seem at first sight unrelated, but all are due to a failure to pump salt across the membranes which surround cells. Medicine has improved the lives of those affected, but few survive beyond their mid-thirties.

Family studies showed long ago that the disease is due to a recessive gene that is not carried on the sex chromosomes. In 1985, pedigrees revealed that it was linked to another DNA sequence which controls a liver enzyme, although it was not then known upon which chromosome that was. Within a year or so, a kindred was discovered in which this pair of genes was linked to a DNA variant that had already been mapped to chromosome seven. The relevant segment of that chromosome was inserted into a mouse cell line, cut into short lengths and the painful task of sequencing begun. By 1988 the crucial region had been tracked down to a segment of DNA one and a half million base-pairs long. Fragments were tested to see if (like the yeast and human sequences later found to control cell division) they had sequences in common with the DNA of other animals as, if they did, the order of letters must have been retained through evolution because they did some unknown but useful job. Several such sections were uncovered. One had an order of DNA letters similar to that of other proteins involved in transport across membranes. It followed the pattern of inheritance of cystic fibrosis. The gene had been tracked down.

The cystic fibrosis gene is a quarter of a million DNA bases long, although the protein has only about one and a half thousand amino acids. Computer models of its shape show that it spans the cell membrane several times, just as expected for a molecule whose job is to act as a pump. Many families with the disease have just one change in the protein: a single amino acid is missing. That changes its shape and stops the new protein from going to the right place in the cell. Instead it is picked up and destroyed by the internal quality-control network.

The discovery of the gene allowed carriers (together with foetuses bearing two copies) to be identified. Unfortunately, cystic fibrosis which once seemed a simple disorder, can, we now know, be caused by many different DNA changes that vary from place to place and from family to family. The illness gave the first hint about the unexpected and unwelcome complexity that the full map was to reveal.

Mapping exploded after that first discovery. At first, the mappers behaved like any explorer in a new territory. A cartographer does not start with a plan of the beach which is then extended in excruciating detail until the whole country is covered. Instead he picks out the major landmarks and leaves the details until later, when he knows what is likely to be interesting. Before today's triumph of technology, most mappers were concerned with a small proportion of the genes, those that lead to inherited disease.

All the most important single-gene inherited illnesses were tracked down within a few years. Huntington's Disease leads to a degeneration of the nervous system and death in middle age. It was once called Huntington's Chorea (a word with the same root as choreography) after the involuntary dancing movements of those afflicted. An eighteenth-century Harvard professor claimed that those with the disease were blasphemers as their gestures were imitations of the movements of Christ on the Cross and some sufferers were burned. It is a dominant, but with a nasty twist: because of the late onset of symptoms, those at risk are left in uncertainty about their predicament. In 1983 came a breakthrough helped by great good luck. Soon after the search started, the approximate site of the Huntington's gene was found by following its association with a linked DNA variant some distance away on the same chromosome. Then, luck ran out, and it took ten years to find the gene. It has now been tracked to the tip of chromosome 4. The shape of the protein which has gone wrong – huntingtin, as it is with some lack of imagination called – has been worked out to give, for the first time, some insight into the nature of the disease, which involves nerve cells in effect committing suicide when the aberrant protein (which looks like nothing else in the cell) instructs them to do so. Many more damaged genes soon fell victim to the genetic explorers and were pinned onto the map.

Type in the four letters OMIM – On Line Inheritance In Man – into any search engine and a list of ten thousand inherited diseases at once appears; symptoms, inheritance patterns, and, for nearly all, chromosomal grid reference. From the hunt for inherited illness, the search shifted to a wider set of genes. No longer were diseases needed as a first clue. To look for genes only when they go wrong is like trying to work out the principles of the internal combustion engine from car breakdowns. Now, the machine itself can be dismantled and its mechanism inferred directly.

When a gene makes something, it generates a complementary molecule – a messenger, as it is known – which transfers information from DNA to the main part of the cell. Because it produces nothing, most DNA generates no messengers at all. To find such molecules is hence an excellent way to search out working genes. There are tens of thousands of distinct messengers. What most do is quite unknown. In most cells, most are switched off but in the brain a large proportion are at work at any time. The brain is more active than is any other tissue (which may help to explain why more than a quarter of all inherited diseases lead to mental illness).

The hunt for genes is more like that for Timbuctu than for El Dorado. The mappers soon found that genes are oases of sense in a desert of nonsense. At one time, it seemed scarcely worth sifting the sands between the genetic cities, but, in the end, the complete map was made mainly on the grounds that it was worth while as one never knows what might turn up. It reaffirmed one of the most mis-understood facts in science; that it is possible to solve most problems by throwing money at them.

The assault on the physical map is best compared to surveying a country with a six-inch ruler, starting at one end and driving on to the opposite frontier. Twenty and more years ago, when the job began, one person could do about five thousand DNA bases a year. Now, it is routine to do thousands of times as many. Much of the intellectual effort of the job has moved from the simple accumulation of information to understanding it. Computer wizardry has played as important a part in the gene map as has biochemical machinery.

Once a segment of DNA has been sequenced, the local maps – the town plans – must be put in the right order. One way to build up a larger chart is to make a series of overlapping sequences of short pieces of DNA. The approach is a little like putting pages ripped out of a street guide back together by looking at the overlaps at the edge of each page in an attempt to find streets which run into each other. Sophisticated programs look for superimposed segments, long or short, and reassemble the torn fragments of DNA. That is much harder than it seems. An alphabet of just four letters and – like the map of an American city

– many repeats of the same pattern of streets, gives plenty of chances for confusion. There are some short cuts. One trick, useful in the early days, was to jump several pages in the guide in the hope of missing out particularly tedious parts of the neighbourhood but for completion even the dullest parts of town must be charted.

New and powerful computers have made it possible, in principle at least, to make a whole genetic atlas at once, rather than piecing it together page by page. The 'random shotgun' approach lives up to its name. It blasts copies of the genome into thousands of segments, again and again, and, like a taxidermist rebuilding a single pheasant from the casual slaughter of many by a blind man with a twelve-bore, reconstitutes the whole thing from scratch. A giant program puts all the shattered pieces together, until at last they look like a map (or a game-bird). That approach worked well in fruit-flies, whose genome was sequenced before that of our own, but flies have a tenth as many DNA letters and far less repetition of easily-confused short sequences than we do. The less audacious 'clone by clone' approach takes tiny fragments (each about a twenty-thousandth of the whole of human DNA) and sequences them one by one. Then, it reassembles short segments of genes and, in time, re-forms the whole atlas. The approach, plodding as it may be, has worked well with humans and was used by the publicly-funded mappers to publish each clone as it appeared and to help thwart the privatised plan to sequence (and patent) the whole of our DNA at one fell swoop.

The physical map does not look at all like the linkage maps which emerged from family studies. The central difficulty is one of scale. A few tens of thousands of functional genes fit into three thousand million DNA letters. As most genes use only the information coded into several thousand bases there seems to be far more DNA than is needed. Mapping shows that just one part in twenty represents part of a gene. Our genome has an extraordinary and quite unexpected structure.

A geographical analogy may help. Imagine the journey along the whole of your own DNA as a trip from Land's End to John o'Groat's via London; about a thousand miles altogether. To fit in all the DNA letters into a road map on this scale, there have to be fifty DNA bases per inch, or about three million per mile. The journey passes through twenty-three counties of different sizes. These administrative divisions, conveniently enough, are the same in number as the twenty-three chromosomes into which human DNA is packaged. With the exception of some short segments a few hundred yards long which, for various technical reasons, have proved recalcitrant, the whole lot has been mapped out with an accuracy of one part in fifty thousand – an inch in a mile (which is as good or better than the maps sold by the Ordnance Survey).

The scenery for most of the trip is tedious. Like much of modern Britain it seems to be unproductive. About a third of the whole distance is covered by repeats of the same message. Fifty miles, more or less, is filled with words of five, six or more letters, repeated next to each other. Many are palindromes. They read the same backwards as forwards, like the obituary of Ferdinand de Lesseps – 'A man, a plan, a canal: Panama!' Some of these 'tandem repeats' are scattered in blocks all over the genome. The position and length of each block varies from person to person. The famous 'genetic fingerprints', the unique inherited signature used in forensic work, depend on variation in the number and position of such segments. Other repeated sequences involve just the two letters, C and A, multiplied thousands of times while yet more are remnants of ancient viruses. Large sections of the genome are given over to long and complicated messages that seem to say nothing.

It is dangerous to dismiss all this DNA as useless because we do not understand what it says. The Chinese term 'Shi' can – apparently – have seventy-three different meanings depending on how it is pronounced. It is possible to construct a sentence such as 'The master is fond of licking lion spittle' just by using 'Shi' again and again. This would seem like empty repetition to those who cannot speak Chinese.

Much of the inherited landscape is littered with the corpses of abandoned genes, sometimes the same one again and again. The DNA sequences of these 'pseudogenes' look rather like that of their functional relatives, but are riddled with decay and no longer make anything. At some time in their history a crucial part of the machinery was damaged. Since then they have been rusting. Oddly enough, the same pseudogenes may turn up at several points along the journey.

After many miles of dull and repetitive DNA terrain, we begin to see places where some product is made. These are the functional genes. They, too, have some surprises in their structure. Each can be recognised by the order of the letters in the DNA alphabet, which start to read in words of three letters written in the genetic code, as a hint that it could produce a protein. In most cases there are few clues about what its product does, although its structure can be deduced (and its shape inferred) from the order of its DNA letters.

Most genes are arranged in groups that make related products, with about a thousand of these 'gene families' altogether. One is involved in the manufacture of the red pigment of the blood. Most of the DNA in the bone-marrow cells which produce the red cells of the blood is switched off. One small group of genes is hard at work. As a result they are better known than any other. Much of human molecular biology grew from research on this particular genetic industrial centre, the globin genes.

They have two factories. One is halfway along the genetic road to John o'Groat's – in Leeds. It makes one part of the protein involved in carrying oxygen. The beta-globin industrial estate contains about half a dozen sections of DNA that code for related things. That responsible for part of adult haemoglobin (and involved, when it goes wrong, in sickle-cell disease) is quite small: about three feet long on this map's scale. A few feet away is another one which makes a globin found in the embryo. Close to that is the decayed hulk of some equipment which stopped working years ago. The beta-globin factory covers about a hundred feet altogether, most of which seems to he unused space between functional genes. It co-operates with a sister estate, the alpha-globin unit, a long way away, (near London, on this mythical map) which produces a related protein. When joined together, the two products make the red blood pigment itself. Most genes are arranged in families, either close together or scattered all over the genome.

The map of ourselves shows that genes are of very different size, from about five hundred letters long to more than two million. One makes the largest known protein, titin, a molecular shock-absorber; a long, pleated structure found in muscles, in blood cells and in chromosomes. Whatever the size of its product, titin is by no means the largest gene. Most human genes have their functional segments interrupted by lengths of non-coding DNA – in Huntington's disease, for example, by nearly seventy In many genes (such as the one which goes wrong in muscular dystrophy) the great majority of the DNA codes for nothing. The non-coding material, whose importance varies greatly from gene to gene, participates in the first part of the production process, but this segment of the genetic alphabet is snipped out of the message before the protein is assembled. This seems an odd way to go about things, but it is the one which evolution has come up with.

The general picture began to emerge as soon as the mappers began work. In the year 2000 – almost exactly a century after the rediscovery of Mendel's rules – their labours were, in effect, complete and the whole human gene sequence was laid out in all its tedium before a less than startled world. Three thousand million letters (or, as now it appears, slightly more) is a lot. For accuracy, each section had to be sequenced ten times or more and even at a thousand DNA bases a second (which is what the machinery pumps out) that was not easy. Sixteen centres, in France, Japan, Germany, China, Britain and the United States combined to do the job. Most were funded by governments of charities, with the notorious exception of the Celera Genomics company (their motto: 'Discovery Can't Wait!'), whose head defected from a government programme. Advances in technology reduced the original estimate of three billion dollars by ten times which, for a project – described by President Clinton as the most wondrous map ever produced – with far more scientific weight than the Moon landings, was a remarkable bargain. For much of the time, the private and public sectors were at daggers drawn (vividly illustrated by Celera's description of the director of one public laboratory behaving as if he had been bitten by a rabid dog).

Because (as so often in science) much of the effort lies not in obtaining information, but in making sense of it, a shotgun marriage between the rivals was, at the last moment, arranged. To 'annotate' the genome – to work out just what the newly-sequenced genes do – was a task so formidable that it demanded the use of one of the world's most powerful supercomputers.

From the mass of data, the small segments that code for

proteins and the even smaller sections that act as the on-off switches for the working genes, are picked out (which is where the computing power comes in). Fortunately, many human genes look rather like those in fruit flies, yeast and nematode worms (all of which have been sequenced) and a massive comparison of each length of human DNA with those in other creatures points at common segments that must, presumably, represent working genes. However, some useless sections disguise themselves as valuable by chance and some pieces have been defined as working genes – again, perhaps wrongly – only by a slight shift in the ratio of particular pairs of bases. As a result, and even with the complete DNA sequence, the precise number of genes needed to make a human being remains, and will remain, uncertain (although most of the researchers guess at a figure of fifty thousand or so).

Even when the functional segments are found, the task of understanding what they do has only just begun. The complete sequence, hailed by Presidents (and Prime Ministers) though it might be, is little more than an arbitrary step on the road to understanding.

Even so, the genome has revealed if not its secrets, at least its structure; and that is remarkable enough. Take, as an instance, chromosome 22; as the smallest of the twenty-three pairs, the Rutland of the genome, and, in the last weeks of the twentieth century, the first to have its entire sequence established. It was mapped with an error rate of fewer than one in fifty thousand bases and just a few short gaps. Apart from its small size (at thirty-three and a half million bases it represents a hundredth of the whole sequence) it is an unremarkable chromosome, quite representative of its larger cousins (the biggest of which, chromosome 1, is eight times longer).

Before the global map, a few scattered genes had been tracked to chromosome 22. They included genes for, among others, a rare disease that causes heart problems and facial distortion, a birth defect called 'cat's eye syndrome' and genes involved in a severe disease of nerve degeneration.

The clone by clone approach – tearing out pages of the map, sequencing each one and ordering them by looking at the overlaps – revealed about a thousand segments of DNA that looked as if they might code for a protein. Seven hundred or so are identical to genes already found within ourselves or elsewhere and may represent useful bits of DNA. Many, no doubt, cause diseases when they go wrong. Some may be impostors; segments that make nothing but have, by chance, an identity that resembles that of a productive section of the genome. Even this, the smallest chromosome, has no shortage of genes. The smallest is a mere thousand bases long, its largest more than five hundred times bigger. Some are uninterrupted, but most are fractured by many inserted sequences of non-coding DNA. To make matters even more complicated, two genes appear within the structure of others. They are genes within genes; read (like a Hebrew sentence in an English book) not from left to right but from right to left.

Most of the remainder of the chromosome is a story of waste and decay. It hides within itself eight or so lengthy duplications, in which whole segments of the instructions are, for some reason, doubled up. The wrecks of genes are everywhere, and about a fifth of the sequences that might once have made something are present only as pseudogenes. Most of the actual workers are – just like the producers of the red blood pigment – members of gene families; and most of those families are filled with genetic black sheep, pseudogenes, who rest on their laurels while their kin go about their business. Some of its genes are responsible for proteins associated with the immune system. They work together as a family of dozens of productive members, but are accompanied by twice as many decayed relatives with a home nearby.

Thirty or so disorders – from cancers to errors in foetal development to a tendency towards schizophrenia – have now been uncovered on this short chromosome: a small part of the thousands that inflict the human race, but a hint of the huge numbers of ways that DNA can go wrong. Most of the proteins mapped to chromosome 22 have no known function. What some do can be guessed at by comparing them with others from elsewhere in the human genome, or from the rest of life, but the majority are anonymous factories, hard at work but with, as yet, no hint about what they made.

Other chromosomes have the same general nature as does the smallest, although each has its quirks. The next in line, chromosome 21, has a personality of its own. Already famous as the sources of the commonest human inborn abnormality, Down's syndrome (present in about one in seven hundred live births) and as the site of one of the genes predisposing to Alzheimer's disease, this structure has some thirty-three million base pairs – about one per cent of the total. One end is stuffed with copies of the same sequence, multiplied again and again; but the other has the machinery. Not, however, very much; for chromosome 21, although about the same size as number 22, has only half its number of genes.

Its two hundred and twenty-five working segments (nine-tenths of which were new to science) seem a modest endowment. Perhaps its depauperised state explains why chromosome 21 is the only chromosome (apart from the sex chromosomes) which the body can tolerate in extra copy. Children with Down's syndrome suffer from fifty or more distinct problems, ranging from heart disease and a tendency to leukemia to difficulties in breathing. One severe problem is their premature ageing and memory loss. That

symptom resembles those of Alzheimer's disease – and one
of the genes responsible for the early-onset form of that
illness is found in chromosome 21. Perhaps some of the
chromosome's other genes will turn out to be associated
with other ailments common to Down's children.

Its other genes include those reponsible for two forms
of inborn deafness and for amyotrophic lateral sclerosis:
a condition known in the United States as Lou Gehrig's
Disease after the New York Yankee slugger who died of
the illness, but in Britain indissolubly linked to the great
physicist Stephen Hawking. Those with the condition suffer from a loss of nerve cells in the brain and spinal cord
and, as a result, slowly lose all power of movement. The
problem lies with an enzyme whose job is to clean up
wastes inside the cell: when it fails, the nerves are slowly
poisoned.

Chromosome 21 may look like a run-down industrial
estate with few of its windows lit up; but at the other
extreme of economic activity, chromosome 6 is full of
active genes. It is in the forefront of the body's genetic
defences. One section has long been known to be responsible for much of the body's immune defences. The crucial
segment is only about a tenth the size of chromosome
21, but contains well over a hundred genes (as well as a
respectable number of relics that have given up the functional ghost). Many share a certain identity and have arisen
by duplication from ancient ancestral genes. Much of their
job involves binding to the proteins of an invader and
passing on information as to its identity to the white blood
cells that then swing into attack. Others code for statements of individual entity on the cell surface.

Because so many genes are involved, there is plenty of
opportunity for reshuffling. Combined with a high rate of
mutation at many of the individual sequences, this generates
a mass of diversity from person to person and from place to

place, as a reminder that to sequence a human genome is only a step in the much larger task of looking for differences among individuals. As some of the genes on chromosome six come in as many as two hundred forms the task will be a lengthy one; but it will be worthwhile because, for reasons unknown, some members of the defence force have the habit of turning on the body itself to give a range of auto-immune diseases such as rheumatoid arthritis, the skin disease psoriasis, and some forms of diabetes.

Whatever the details of its individual chromosomes, the picture to emerge from the completed human genome project is one of size, and of complexity. Like the New World that stretched before the first explorers, the genome is, above all, big; and like all large objects has a gravitational attraction for metaphors. Printed out, the information gathered by the Human Genome Project will fill two hundred telephone books or, in a single line, stretch the length of the Mississippi; a secretary would take twenty years to type (and a cantor fifty to sing) the whole thing, and so, symbolically, on.

Through all the symbolism and hyperbole is emerging an uneasy feeling that all this has been a race only to the starting line of understanding our genes and not to the winning post. The journal *Nature* reported the analysis of one of the very simplest viral genomes in the 1970s under the headline 'Sequencing is Not Enough'. That message emerges with renewed strength from the human genome project.

Even so, its completion is a milestone in the history of genetics. To look at an ancient chart – even one as faulty as that of Herodotus – is to realise that maps contain within themselves a great deal about the lives of those who drew them. They show the size and position of cities, the paths of migration, and the record of peoples long gone. The chart of the genes is made; and now the real journey can begin.

Chapter Four

CHANGE OR DECAY

By the time you have finished this chapter you will be a different person. I do not mean by this that your views about existence – or even about genes – will alter, although perhaps they may. What I have in mind is simpler. In the next half hour or so your genes, and your life, will be altered by mutation; by errors in your own genetic message. Mutation – change – happens all the time, within ourselves and over the generations. We are constantly corrupted by it; but biology provides an escape from the inevitability of genetic decline.

Evolution is no more than the perpetuation of error. It means that progress can emerge from decay. Mutation is at the heart of human experience, of old age and death but also of sex and of rebirth. All religions share the idea that humanity is a decayed remnant of what was once perfect and that it must be returned to a higher plane by salvation, by starting again from scratch. Mutation embodies what faith demands: each man's decline but mankind's redemption.

The first genes appeared some four thousand million years ago as short strings of molecules which could make rough copies of themselves. At a reckless guess, the original molecule in life's first course, the primeval soup, has passed through four thousand million ancestors before ending up in you or me (or in a chimp or a bacterium). Every one of the untold billions of genes that has existed since then emerged through the process of mutation. A short message has grown to an instruction manual of three thousand million letters. Everyone has a unique edition of the instruction book that differs in millions of ways from that of their fellows. All this comes from the accumulation of errors in an inherited message.

Like random changes to a watch some of these accidents are harmful. But most have no effect and a few may even be useful. Every inherited disease is due to mutation. Now that medicine has, in the western world at least, almost conquered infection, mutation has become more important. About one child in forty born in Britain has an inborn error of some kind and about a third of all hospital admissions of young children involve a genetic disease. Some damage descends from changes which happened long ago while others are mistakes in the sperm or egg of the parents themselves. Everyone carries single copies of damaged genes which, if two copies were present, would kill. As a result, everyone has at least one mutated skeleton in their genetical cupboard.

Because there are so many different genes the chance of seeing a new genetic accident in one of them is small. Even so, in a few cases, novel errors can be spotted.

Before Queen Victoria, the genetic disease haemophilia (a failure of the blood to clot) had never been seen in the British royal family. Several of her descendants have suffered from it. The biochemical mistake probably took place in the august testicles of her father, Edward, Duke of Kent. The haemophilia gene is on the X chromosome, so that to be a haemophiliac a male needs to inherit just one copy of the gene while a female needs two. The disease is hence much more common among boys. This was known to the Jews three thousand years ago. A mother was allowed not to circumcise her son if his older brother had bled badly at the operation and, more remarkably, if her sister's sons had the same problem.

As well as its obvious effects after a cut, haemophilia does more subtle damage. Affected children often have many bruises and may suffer from internal bleeding which can damage joints and may be fatal. Once, more than half the affected boys died before the age of five. Injection of the clotting factor restores a more or less normal life.

Several of Victoria's grandsons were haemophiliacs, as was one of her sons, Leopold. Two of her daughters – Beatrice and Alice – must have been carriers. The Queen herself said that 'our poor family seems persecuted by this disease, the worst I know'. The most famous sufferer was Alexis, the son of Tsar Nicholas of Russia and Queen Alexandra, Victoria's granddaughter. One reason for Rasputin's malign influence on the Russian court was his ability to calm the unfortunate Alexis. The gene has disappeared from the British royal line, and no haemophiliacs are known among the three hundred descendants of Queen Victoria alive today. In Britain, about one male in five thousand is affected.

Somewhat incidentally, another monarch, George III, may have carried a different mutation. The gene responsible for porphyria can lead to mental illness and might have been responsible for his well-known madness. The retrospective diagnosis was made from the notes of the King's physician, who noticed that the royal urine had the purple 'port-wine' colour characteristic of the disease. A distant descendant also showed signs of the illness. One of the King's less successful appointments was that of his Prime Minister, Lord North, who was largely responsible for the loss of the American Colonies. It is odd to reflect that both the Russian and the American Revolutions may in part have resulted from accidents to royal DNA.

Research on human mutation once involved frustration ameliorated by anecdotes like these. It has been turned on its head by the advance of molecular biology. In the old

days, the 1980s, the only way to study it was to find a
patient with an inherited disease and to try to work out
what had gone wrong in the protein. The change in the
DNA was quite unknown. This was as true for haemophilia as for any other gene. In fact, haemophilia seemed
a rather simple error. Different patients showed rather different symptoms, but the mode of inheritance was simple
and all seemed to share the same disease.

Now whole sections of DNA from normal and haemophiliac families can be compared to show what has happened and, like the genetic map itself, things have got more
complicated. Molecular biology has made geneticists' lives
much less straightforward. First, uncontrollable bleeding
is not one disease, but several. To make a clot is a complicated business that involves several steps. Proteins are
arranged in a cascade which responds to the damage, produces and then mobilises the material needed and
assembles it into a barrier. A dozen or more different genes
scattered all over the DNA take part in the production
line.

Two are particularly likely to go wrong. One makes
factor VIII in the clotting cascade. Errors in that gene lead
to haemophilia A, which accounts for nine tenths of all
cases of the disease. The other common type – haemophilia
B – involves factor IX. In a rare form of the illness factor
VII is at fault.

Factor VIII is a protein of two thousand two hundred
and thirty-two amino acids, with a gene larger than most
– about 186,000 DNA bases long, which, on the scale
from Land's End to John o'Groat's, makes it about a hundred yards long. Just a twentieth of its DNA codes for
protein. The gene is divided into dozens of different functional sections separated by segments of uninformative
sequence. Much of this extraneous material consists of
multiple copies of the same two-letter message, a 'CA

repeat'. There is even a 'gene-within-a-gene' (which produces something quite different) in the factor VIII machinery.

The haemophilia A mutation, which once appeared to be a simple change, is in fact complicated. All kinds of mistakes can happen. Nearly a thousand different errors have been found. Their virulence depends on what has gone wrong. Sometimes, just one important letter in the functional part of the structure has changed; usually a different letter in different haemophiliacs. The bits of the machinery which join the working pieces of the product together are very susceptible to accidents of this kind. In more than a third of all patients part, or even the whole, of the factor VIII region has disappeared. A few haemophiliacs have suffered from the insertion of an extra length of DNA into the machinery which has hopped in from elsewhere.

Once, the only way to measure the rate of new mutations to haemophilia (or any other inherited illness) was to count the sufferers, estimate the damage done to their chances of passing on the error and work out from this how often it must happen. Technology has changed everything. Now it is possible to compare the genes of haemophiliac boys with those of parents and grandparents to see when the mutation took place.

If the mother of such a boy already has the haemophilia mutation on one of her two X chromosomes, then she must herself have inherited it and the damage must have occurred at some time in the past. If she has not, then her son's new genetic accident happened when the egg from which he developed was formed within her own body. In a survey of a British families with sons with haemophilia B (whose gene, that for Factor IX, is 33,000 bases long) many different mutations were found, most unique to one family. Eighty per cent of the mothers of affected boys had themselves inherited a mutation. However, in most cases the damaged gene was not present in their own father (the grandfather of the patient). In other words, the error in the DNA must have taken place when his grandparental sperm was being formed.

A quick calculation of the number of new mutations against the size of the British population gives a rate for the haemophilia B gene of about eight in a million. The difference in the incidence of changes between grandfathers and their daughters suggest that the rate is nine times higher in males than in females. The sex difference is easy to explain. There are many more chances for things to go wrong in men (who – unlike women – produce their sex cells throughout life, rather than making a store of them early on, and hence have many more DNA replications in the germ line than do females). For some genes the rate of mutation among males is fifty times higher than in the opposite sex. Men, it seems, are the source of most of evolution's raw material.

Most people with severe forms of haemophilia have each suffered a different genetic error. Such mistakes happen in a parent's sex cells and disappear at once because the child dies young. Those with milder disease often share the same change in their DNA; an error that took place long ago and has spread to many people. The shared mutation is a clue that these individuals descend from a common ancestor. The non-functional DNA in and around the haemophilia gene is full of changes which appear to have no effect at all and have passed down through hundreds of generations. Near the gene itself is a region with many repeats of the same message. The number of copies often goes up and down, but its high error rate seems to do no damage.

All this hints that mutation is an active process, with plenty of churning round within the DNA. This new fluidity once alarmed geneticists as it violates the idea of gene as particle (admittedly a particle which sometimes makes mistakes) which used to be central to their lives. So powerful is the legacy of Mendel that his followers have sometimes been reluctant to accept results which do not fit. This is very true of some of the new and bizarre aspects of mutation.

Scientists, in general, despise doctors. For many years, physicians reported a strange genetical effect called 'anticipation'. The malign effects of some inherited diseases seemed to show themselves at a younger age with each generation that passed. The effect was named eighty years ago by an enthusiastic eugenical doctor called Mott. He thought that it presaged the inevitable degeneration of society: 'The law of anticipation of the insane represents . . . rotten twigs continually dropping off the tree of life.' Later geneticists were resistant to the idea and it disappeared from view. In fact it represents a new kind of mutation, an inherited error which gets worse as the generations succeed and which is now known to be common.

The process is seen in a disease called the fragile X syndrome, the most important single cause of inborn mental impairment, with symptoms that range from mild to crippling. Many children diagnosed as autistic have in fact a minor form of this illness. At first sight its inheritance is odd, as in some families it is found in just one person, whereas others have dozens of affected members. Boys tend to suffer more damage than do girls, with mental retardation and, sometimes, a characteristic face with large ears, and heart problems.

Although both males and females are affected, the gene is sex-linked. Males never pass it on to their sons, but girls whose mothers have it are carriers, and some may be affected. Fragile-X is one of the few genetic diseases in which the damage can be seen down the microscope, for near the end of every affected X chromosome is a small constriction which looks as if it might be about to break. About one woman in two hundred and fifty has one or other X chromosome damaged in this way. Many show no symptoms at all and neither do their children. Others have signs of the disease, as do their offspring, while a fraction albeit themselves normal may have children with the illness.

The mutation is a multiplication of a three-letter DNA repeat – C, G and G – within a gene. Its protein helps form the connections made as the young brain begins to respond to experiences from the outside world. The damage is not in the coding section of the gene, but in its on-off switch. The mutation is flexible. Most people have thirty or fewer repeats, and some have as few as half a dozen. When the number creeps above fifty or so, children are in danger and may find it hard to speak or to read and as it rises over two hundred (and severely affected individuals have more than a thousand repeats) the full symptoms set in.

The strangest aspect of fragile X – and, we now know, of many other mutations – is that the number of repeats (and the amount of damage) changes from generation to generation. The daughter of a mother who has fragile X is more likely to have an affected child than was her own parent although (or so it seems) she has passed on exactly the same gene. Each generation, the number of copies changes, going up when it is transmitted through a female, but staying the same or decreasing when a man passes on the damaged chromosome.

One form of muscular dystrophy also shows more virulent effects as the generations succeed. Again, a repeated sequence is involved. The pedigrees of one group of the children with the disease shows that all shared an ancestor who lived in the seventeenth century. He was healthy; as were, for two hundred years, his descendants; but suddenly some, distant relatives though they now are, began to suffer from dystrophy. More copies of a DNA repeat within the crucial gene had been made each generation. Once a critical number was reached the symptoms appear. Each generation, more and more appear, and the effects of the damaged gene become more severe as it passes down the lineage. Huntington's Disease, too, is due to a repeat of the three DNA letters CAG. Each triplet codes for a single amino acid, which shoulders itself into the middle of the huntingtin molecule. Some people have fewer than ten copies, some more than a hundred. Once more than about thirty-five copies are made, the symptoms of the disease emerge, and the more repeats, the earlier they do so. Those with fifty are in danger of illness while still in their twenties. Half a dozen other diseases of the nervous and muscular system are due to such three-letter intruders. Why nerves should be so prone to them is not certain, although the tendency of such augmented proteins to form great clumps in the cell may have something to do with it.

If the rate of mutation to haemophilia is taken as the norm, there must be about one new DNA change in a functional gene per five generations. This means ten million changes in functional genes per generation in Britain. The actual incidence may be even higher. Hormonal changes in women who are attempting to become pregnant show that eight out of ten fertilised eggs are lost. Many may carry new lethal mutations. Often, they involve the loss of all or part of a chromosome, and the incidence of such errors in still-born children is ten times that among those born alive.

Each gene has its own mutation rate. The frequency varies more than a thousand times from gene to gene. Larger genes with more interspersed pieces of DNA go wrong more often than smaller ones, and certain combinations of bases change more readily than others. The short segments of repeated DNA outside the functional genes (such as those involved in the 'genetic fingerprint') have a high rate of error. As many as one person in ten may pass on a change. The rate of mutation itself has evolved, too, and is controlled by enzymes which can repair injured DNA. When these are themselves damaged it shoots up.

Many things increase the mutation rate. Radiation, for example, can have a powerful effect. Plenty of mutations do not arise from the natural instability of the genes, but from damage inflicted from outside (as many of the early workers on radium, many of whom died of cancer, soon found out). Up to two thirds of the sperm cells of cancer patients who have been given large X-ray doses carry chromosomal changes. Evidence from other animals makes lower doses of radiation a real cause for concern, given the link between agents that cause mutations in sperm and egg and those that cause cancer. The acceptable dose for humans is set in part by research on mice (which seem to be more susceptible than we ourselves are) and there have been calls to have the limits increased; but nobody denies that radiation damages our genes. Sunlight, too, is harmful to cells and even an increase in temperature can increase the rate of error.

The biggest avoidable source of radiation in Britain is radon gas, which leaks from granite. People who live in granite houses in Cornwall may be exposed to more excess radiation than are those who work in nuclear power stations (although their equivalents in the granite city of Aberdeen may in part reverse the effect as they wear the kilt, with its cooling influence). In the United States, houses built with radioactive sands in their foundations have been demolished as their occupants faced twenty times the average dose. In the UK, those at risk are advised to install fans to stop the build-up of gas. Other sources of radiation include the cosmic rays experienced during air travel and medical X-rays, but for most people these involve very small doses.

Chemicals are much more important agents of genetic damage. The number of chromosome errors in nuclear power-station workers is a little greater than that of the general public, but the number in those employed in coal-fired stations is even higher because of the noxious byproducts of burning coal. Bacteria are used to test a huge number of likely, and some unlikely, substances. Some, such as those once used in hair dyes, had a powerful effect and have been banned. Others, those in black pepper, in Earl Grey tea and in some pesticides, also cause mutations. Some of the most potent are quite natural. Plants produce many toxic chemicals for defence against insects and even lettuce, the epitome of a healthy diet, contains substances that cause mutations in mice. Almost half of all cancers may be influenced by the food we eat; and the vast majority of the pesticides – perhaps more than 99.9% – in the Western diet are perfectly natural. Cynics argue that organic foods are more dangerous than food which has been sprayed because of the noxious chemicals found in the moulds which grow on them. Fresh fruits and vegetables reduce the rate, and some plants, such as broccoli and tomatoes, are filled with anti-mutagens that may help protect those who eat them.

Mutations are the raw material of evolution. Life progresses; it does not decay, but every individual is mortal. As we grow old our machinery corrodes until at last it breaks down.

Part of this erosion comes from genetic changes within the body and part from the delayed effects of genes advantageous when young but harmful when old. To build an adult from a fertilised egg involves making hundreds of millions of cells, each with its own copy of the original message. The copying process is imperfect, and there are plenty of chances for mistakes. Even in adulthood most cells continue to divide. Red blood cells, for example, are renewed every four months or so. Every minute everyone makes thousands of miles of DNA. As a result, huge numbers of mutations build up in body cells. Each individual is an evolving system whose identity changes from day to day.

Some of these changes can lead to disaster. Many cancers result from genetic accidents. Indeed, some cancers look more and more like genetic diseases. They represent a decay of the genetic message and a loss of control by DNA of the cells in which it lives. Age is a reflection of the same process. As our bodies are in a constant fever of replication, the older we are the more divisions there have been and the more chance for error. The cells of a new-born baby are separated by just a few hundred divisions from the egg; but mine, as a fifty-something-year-old are distanced from it by thousands. My genes have had more chances to mutate than have those of a baby. What is worse, they are less effective at repairing the damage. The cells of old people even contain altered genes which make inappropriate proteins. Thus, many aged Europeans have small amounts of sickle-cell haemoglobin in their blood. This gene is normally found in Africans but has, in their case, appeared as a new mutation within their elderly bodies.

Ageing accelerates with age. The lowest risk of death is at about the age of twelve – just before puberty. After that, the rate doubles every eight years or so, giving a seventy-six-year old about a two hundred and fifty times greater chance of dying than a teenager. The power of accelerated decay is impressive. If the death rate stayed at that of a twelve-year-old, most people would live to a thousand and there would be a small but noticeable proportion of people around who were born in the last Ice Age. Unfortunately, our obsolescence is such that even centenarians are rare. All this helps to explain why cancer is a disease of the old; so much so that even if the disease were eliminated altogether life expectancy would go up by only about four years. The biological identity crisis which we define as old age and which is solved by death happens when the genetic message becomes so degenerate that its instructions no longer make sense. The rate of ageing is programmed. Mouse cells in culture stop dividing after about four years, while human cells can carry on for almost a century.

Parts of the message disappear with time. DNA is packaged into chromosomes. Each has a specialised length of DNA at its end that marks the point at which the DNA-duplication machinery stops and loops back on itself, rather like the crimp at the ends of shoelaces that stops them from fraying. This gets shorter with age. In a baby it is about twenty thousand letters long, while in a sixty year-old it is less than half that length. Cells from tumours have lost even more DNA from the chromosome ends. About forty letters are dropped from this section of the message each time a cell divides, so that an old body works from an imperfect instruction manual, full of typographic errors. The same happens to mitochondrial genes, which are shot full of holes as the years continue their inexorable progress.

Old age is itself in part the result of genetic accidents. Human cells in culture age more quickly when they carry a defect which increases the mutation rate and some children who inherit a tendency towards cancer also show symptoms of senility much earlier than normal. The immune system, which has the highest mutation rate of any part of the body, often fails as the years pass by. It seems that the decay of our elderly selves is, to some degree, a consequence of mutation. The influence of old age in damaging sperm and egg adds a certain irony to the claims of one institution devoted to the reversal of the decay of the human race, the Centre for Germinal Choice in California, in which Nobel Prize-winners make genetic deposits for hopeful mothers. The depositors may once, as they claim, have approached a genetic ideal, but that perfection has been marred by age.

Why, if our genes change and decay through our lives, does the human race not degenerate as one generation succeeds another? The answer lies in sex. To define sex is simple; it is a process that brings together genes from different ancestors. It provides a chance to purge ourselves of the harmful mutations which arise in each generation and represents, in more ways than one, the antithesis of age.

Almost every novel, play or work of art revolves around the eternal triangle of sex, age and death. All three – and our very existence – emerge from errors in the transmission of genes. Humanity is not a degenerate remnant of a noble ancestor. Rather we are the products of evolution, a set of successful mistakes. Genetics has solved one of our oldest questions; why people decay, but *Homo sapiens* does not.

Chapter Five

CALIBAN'S REVENGE

The plot of George Eliot's novel *Daniel Deronda* is a convoluted one. It revolves around the adventures of Daniel himself, the adopted son of a baronet. After some hundreds of pages he develops an unexpected interest in things Hebrew and – some time later – it transpires that Daniel Deronda was, quite unaware, the son of a Jewish woman. His biology had triumphed over his background.

Many people are obsessed by the role of inheritance compared to that of experience. The infatuation goes back long before genetics. Even Shakespeare had a say: in *The Tempest* Prospero describes Caliban as 'A devil, a born devil, on whose nature Nurture can never stick.' There are still endless (and rather empty) discussions about whether musicality, criminality or intelligence is inherited or acquired and more serious debates about the role of genes and environment in illnesses such as cancer or heart disease. Such questions are often unresolved, and may be unresolvable.

Galton, in *Hereditary Genius*, went to great lengths to show that talent runs in families and was coded into their biology. He failed to point out that more than half his 'geniuses' turned up in families with no history of distinction at all and concentrated only on those who supported his hereditarian views. Most claims that talent (or lack of it) is inherited are based, like Galton's, on little more than a series of selected anecdotes. Even the descendants of Johann Sebastian Bach disappeared from the musical firmament after a few generations. Family likeness says little about the importance of biology; after all, one attribute much shared by parents and children is bank-balance.

Nevertheless, the question of nature versus nurture is of endless fascination. Dozens of studies purport to show that behaviour is under genetic control. Whole sets of degenerate families were once held up for inspection: the Tribe of Ishmael, the Jukes Clan and the Kalikaks (whose pseudonym is Greek doggerel for good/bad). One was traced to an eighteenth century sailor who married an upright woman but had an affair with a slattern. His wife's branch gave rise to a lineage of spotless virtue while the other was a burden on society, as firm proof that morality lies in the genes.

Geneticists find queries about the importance of nature and nurture dull, for two reasons. First, they scarcely understand the inheritance of complex characters (those, like height, weight or behaviour which are measured rather than counted) even in simpler beings like flies or mice and even with traits which are easy to define. Second, and more important, geneticists know that the perpetual interrogation – gene or environment? – is often meaningless. Its only answer is that there is no valid question.

Although genetics is all about inheritance, inheritance is certainly not all about genetics. Almost every attribute involves the joint action of the internal and the external world. A characteristic such as intelligence (or height) is often seen as a cake ready to be sliced into so much 'nature' and so much 'nurture'. In fact, the two are so closely blended that to separate them is like trying to unbake the cake. A failure to understand this simple fact leads to confusion and worse.

Not far from Herbert Spencer's (and his neighbour Marx's) tomb is a large red-brick house. It was occupied by Sigmund Freud after he fled Austria to avoid racial policies which descended from the Galtonian ideal. On his desk is a set of stone axes and ancient figurines. Freud's interest in these lay in his belief that behaviour is controlled by biological history. Everyone, he thought, recapitulates in their childhood the phases experienced during evolution. Freud saw unhappiness as a sort of living fossil, the re-appearance of ancient behaviour which is inappropriate today. Like Galton he saw the human condition as formed by inheritance. The libido and ego are, he wrote, 'at bottom heritages, abbreviated recapitulations of the development which all mankind passed through from its primaeval days'. Freud hoped that once he had uncovered the inborn fault which caused despair, he might be able to cure it.

Today's Freudians have moved away from their guru's Galtonising of behaviour. They feel that nurture is important. Analysis looks for childhood events rather than race memories. In so doing they are in as much danger as their master of trying to unbake the cake of human nature. Any attempt to do so is futile.

The Siamese cat shows how pointless the task may be. Siamese have black fur on the tips of the ears, the tail and the feet, but are white or light brown elsewhere. They carry the 'Himalayan' mutation, which is also found in rabbits and guinea pigs (but not, alas, in humans). Crosses show that a single gene that follows Mendel's laws is involved. At first sight, then, the Siamese cat's fur is set in its nature: if coat colour is controlled by just one gene then surely there is no room for nurture to play a part.

However, the Himalayan mutation is odd. The damaged gene cannot produce pigment at normal body temperature but works perfectly if it is kept cool. As a result, the colder parts of the cat's body, its ears, nose and tail (and, for a male, its testicles) are darker than the rest. An unusually dark cat can be produced by keeping a typical Siamese in the cold and a light one by raising it in a warm room. Inside every Siamese is a black cat struggling to get out. To ask whether its pattern is due to gene or to environment means nothing. It results from both. What the Siamese cat – and every other creature – inherits is an ability to respond to the circumstances in which it is placed.

Many inborn diseases show this effect. The recessive abnormality phenylketonuria (or PKU) affects about forty British children a year. Each has an inherited defect in a particular enzyme which means that they cannot process an amino acid, phenylalanine, found in most foods. As a result they build up large amounts of a harmful by-product. Untreated, such children have low intelligence and die young. The fate of those with PKU is, it seems, sealed by their genes.

But most PKU children born today lead more or less ordinary lives. A change in the environment saves them. If they are diagnosed early (and all babies are tested at birth), they can be given food which lacks all but a tiny amount of phenylalanine. They then develop as healthy infants. Their nature has been determined by careful nurturing and the question of whether DNA or diet is more important to their health has no answer.

Hundreds of genes show the same interaction. A whole new science turns on individual differences in the response to drugs. The genes involved were unknown until humans began to manipulate their chemical milieu. A few people carry an inherited variant which makes them fatally sensitive to a muscle-relaxant used before surgery and everyone is now tested to see whether they are at risk before the drug is given. One of the stranger injunctions of Pythagoras was a caution to his followers not to eat broad beans. He died because, pursued by a mob enraged by his philosophical views, he refused to escape across a beanfield. Pythagoras lived in the Italian city of Croton. Many of its modern inhabitants feel unwell if they eat partly cooked beans. One of the side-effects of the thalassaemia gene (which is common there) is to remove the ability to break down a chemical found in broad beans (and another one used as an anti-malarial drug). When gene and bean (or drug) are brought together, the results can be unpleasant or, in the case of the drug, worse.

All this means that the boundaries between inherited disease and what was is governed by the external world have become blurred. That alters the way we think about medicine. Individual treatments may soon be tailored to a patient's biological heritage. Two disorders, anencephaly and spina bifida, cause a failure of development of the spinal cord; and each runs in families. Part of the problem, though, has to do with poor diet. Their incidence shot up in Holland after the famine of 1945 and both are frequent in Ireland and in Scotland (places known for an unhealthy diet). Mothers who have had an affected child now take vitamin supplements in later pregnancies. This reduces the chance of their genes damaging their children.

A change in the environment can also cause genetic disease. Hay fever was not recognised as a distinct illness until 1819, when it (and its relatives asthma and eczema) were seen as afflictions of the rich. Now, about half the people of the western world are, or claim to be, allergic to one substance or another. In Britain one child in four has asthma. The lung becomes inflamed and its muscles sensitive to the slightest irritation. The unfortunate patient wheezes and coughs, and may suffer permanent damage – and, sometimes, even sudden death. The illness involves an over-reaction by the immune system to an external stimulus. House-dust mites are one culprit, cats another, pollen a third. They were around before 1819, but, for some reason, caused few problems.

Part of the reason lies in the modern world, with its obsession with cleanliness. This may have abolished many infectious diseases, but allows others, once rare, to reveal our inborn weaknesses. Asthma is a disease of the middle class; more common in those well fed as children, in infants dosed with antibiotics, and in Western rather than Eastern Europe. The children of farmers and of those with dogs have less chance of the illness than do vegetarians in a pet-free home. It is an affliction of Thrushcross Grange rather than of Wuthering Heights.

Emily Brontë knew the answer. Cathy, when she returns clean and demure after her convalescence at Thrushcross Grange is faced by Heathcliff's: 'I shall be as dirty as I please; and I like to be dirty, and I will be dirty!' Filth is the key. Infants born into clean households are deprived of an essential learning experience. Not their brain but their immune system lacks stimulation. Middle-class homes lack the grime with which humankind evolved. The immune system, like the brain itself, must be trained to deal with the challenges that it will face later in life. Each needs stimulation; but the immune system demands tapeworms rather than Mozart.

Whatever the importance of dirt, asthma and its relatives have an inherited component. Identical twins are more likely each to suffer than are non-identicals, and those who bear certain variants in genes that code for elements of the immune system are also liable to become ill. Tristan da Cunha, that distant and inbred island in which many people share the same genes, has an epidemic of asthma, with almost half the population affected.

Allergy is a classic of the interaction of gene and environment. Long ago, the genes that today cause problems may have been useful as those with an active immune system were good at resisting infection. After soap, in an unnaturally clean household, an over-active immune system became a nuisance as it disposes to allergy. The environment has changed, but the genes remain the same. Today's DNA has quite different effects on health than it once did, and a change in the interaction of nature with nurture leads to an outbreak of illness.

The term 'cancer' covers a multitude of conditions. All are due to a failure to control cell division. Hundreds of genes control the growth of cells and, when they mutate, the process may go out of control. As in haemophilia, all kinds of mistakes can happen. A single DNA base may change or whole sections of the message be lost. Sometimes the error involves genes moving from one chromosome to another, or from the effects of viruses. Often, several different genetic accidents are needed to promote the development of a tumour. The general picture is not much different from that of mutation in sperm or egg.

Cancer is a Siamese cat of an illness, and the chances of contracting it depend both on the genes and the circumstances with which they are faced. Cell division needs brakes and accelerators. The first, tumour suppressor genes as they are called, control a set of proto-oncogenes that encourage cells to grow. If either party goes wrong, then division speeds up. The cell, though, has a set of speed cameras that control rogue genes. It must pass through a number of checkpoints on the road to division, and if anything is suspicious the cell dies (which is, after all, the natural fate of most cells). Many of the causes of cancer increase the amount of a specific protein – p53, as it is called – that is sensitive to any sign of DNA damage in the cells under attack. Most then commit suicide rather than causing trouble. Indeed, many cancer treatments (themselves often agents of the illness, like radiation and certain chemicals) themselves wake up the p53 genes and persuade the cells to do the decent thing. Damage to the checkpoint itself (either inherited, or caused by the external agent) means real trouble: by that time the rogue line of cells is through the last safety barrier and may be imposs-ible to contain.

Some cancers are more common among those exposed to a particular hazard. Many chimney-sweeps died of a skin cancer, which appeared first on the scrotum. The Eng-lish physician Percival Pott suggested that soot was blame. He was right. Soot, oil and tar contain many carcinogenic chemicals. Radiation, too, can be dangerous. As many as two thousand cases of lung cancer per year in Britain – a twentieth of the total – arise from exposure to radon. There were once thought to be clusters of childhood leukaemia cases around nuclear power-stations but these have now been dismissed on statistical grounds. For most Britons, exposure to radiation is so low that it cannot be an impor-tant general cause of cancer.

Some may be less fortunate. The Techa River runs through Chelyabinsk, once the nuclear capital of Russia. During the height of the Cold War, so much waste was dumped that a fisherman on its banks could get a lethal dose in a week. Many riverside villages have been moved, but the inhabitants of those that remain have increased levels of leukaemia and of cancers of the thyroid and other organs. Even so, the effects are less than those suffered by the survivors of the Hiroshima bomb, so that a sudden burst of radiation may be more dangerous than the same dose given over a longer period. As western safety stan-dards are based on the Japanese cancer figures, there have been calls to relax the minimum dose, but those are resisted by many who believe that no safe lower limit exists.

Whatever the effects of radiation on cancer, chemicals are more important. Those in tobacco smoke are potent agents and some industrial chemicals are just as bad. Alco-hol, too, is far from blameless. Certain chemicals bind to DNA to cause their damage. The amount of bound material gives an estimate of exposure to mutagens. The Polish city of Gliwice, which burns much soft coal, is one of the most polluted places in the world. Gliwice has a high rate of cancer. Many inhabitants have large amounts of poisonous chemicals stuck to their DNA. The amount goes up in the winter, when the smoke is at its worst. Many of those exposed will develop the disease.

Other cancers (such as retinoblastoma, a degenerative disease of the retina) run in families, with no obvious environmental link. The causes of the disease run all the way from gene (predominant in retinoblastoma) to environment (equally so in scrotal cancer), but usually includes both. Workers in the primitive oil industry believed that people with fair hair and freckles should not be employed they were at risk of 'sootwort', as scrotal cancer was known. As such people are in more danger of skin cancer when exposed to sunlight, there may be some truth in the idea.

Seven million die of cancer each year. Many expose themselves to environments so dangerous that even the finest genes cannot save them. Caliban himself could not have devised a contrivance as fiendish as the cigarette: a cheap drug delivery system that provides a narcotic as addictive as heroin and some of the most carcinogenic of all chemicals. Hundreds of millions of people have volunteered themselves as subjects in a gigantic experiment and millions have obligingly died. Their generosity proves the joint actions of nature and nurture. If everybody smoked, lung cancer would be a genetic disease. Many of the cellular checks and balances have a lot of natural variation. For one crucial member, about one person in ten has a highly active form which, when faced with tobacco smoke, does its job very well – and, as it does so, produces a dangerous carcinogen. As a result, light smokers with this form of the gene face a seven times greater risk of lung cancer than do those with other variants (although in heavy smokers, who batter all their defences into submission with massive doses of poison, the risk merely doubles).

Even when DNA is damaged, it can be repaired. A few families lack one or other of the enzyme systems involved and as a result are at high risk. Their activity in the population as a whole varies a hundredfold; and, once again, those with the feebler forms are at increased risk if they smoke. Indeed, such forms are five times commoner among smokers with lung cancer than in those who escape the disease. Blacks who smoke have higher rates than do whites, and this too is associated with some unidentified genetic difference among the groups.

Smokers can choose whether to poison themselves but others are not so lucky and fall, through no fault of their own, into the pit dug by their genes. Liver cancer is the fifth commonest cancer in the world with its capital in Africa and in China. The immediate cause is aflatoxin; a chemical made by the moulds that grow on badly stored foods such as peanuts, rice, beans and other staples of the tropical diet. They destroy the immune system, stunt growth, and cause cancer. Those with the disease have a new mutation in a gene whose normal role is to prevent cells from uncontrolled division. The mutation is of the kind produced by aflatoxin in the laboratory and the peoples of these regions have high levels of the poison in their blood. Improvements in food storage could control liver cancer. Poverty means that even this may not be achieved.

Moves are afoot to protect at least some of those at risk. Heavy smokers – about one in ten of whom will develop lung cancer – are given vitamin A in the hope of reducing the effects of mutations in lung cells. Those who inherit a gene that predisposes to colon cancer are treated with aspirin before they develop symptoms as this might reduce its effects. Such illnesses are sometimes seen as a kind of programmmed Nemesis about which nothing can be done. An appreciation of the role of the environment shows this to be untrue.

Disorders such as cancer and heart disease do run in families but their inheritance is hard to study. Many genes are involved and the circumstances faced by those at risk play a part. One way of exploring them is to use twins, nature's own experiment in human genetics.

Twins are of two kinds, identical and non-identical. Non-identical twins come from the fertilisation of two eggs by two sperm (and now and again turn out to have different fathers). Such twins have half their genes in common and are no more similar than are brothers or sisters. Their situation is – yet again – described in that fount of early genetics, the Old Testament. Jacob and Esau were twins; but 'Esau was a cunning hunter, a man of the fields; and Jacob was a plain man, dwelling in tents'. They looked quite different – 'Behold, Esau my brother is an hairy man, and I am a smooth man' – and even had different manners of speech: 'The voice is Jacob's voice, but the hands are the hands of Esau.'

Such twins are not uncommon. In marmoset monkeys most-births are of this kind. For no obvious reason, their numbers vary from place to place. In Europe, about eight births per thousand are of fraternal twins. France has rather fewer and Spain rather more than the average. Among the Yoruba, in Nigeria, the figure is five times higher. Older mothers tend to have more twins, as do those who have already had several children.

Identical twins are rarer, at about four per thousand births, a rate which does not change much from place to place. In few mammals are they common, but the armadillo always gives birth to identical quadruplets. Identical twins result from the division of an egg which has already been fertilised. They share all their genes and have long been a source of legend. Castor and Pollux, the heavenly twins, were identical as were their equivalents in Germanic legend, Baldur and Hodur (not to speak of Romulus and Remus, the founders of Rome).

Twins can be used in several ways to study nature and nurture. The simplest (but by far the least common) is to find identical twins separated at birth and brought up in different households. If a character is under genetic control the twins should be the same in spite of their contrary circumstances. If environment is more important, each twin should grow to resemble the family with which they spent their childhood.

This simple plot is the basis of a great deal of fiction, in science as much as in literature. Many studies have claimed to show that identical twins reared apart were similar in size, weight or sexual orientation, but much of this work was unreliable. Often, the adoptive families were similar in social position, or the twins knew each other as they grew up. Twins who believed themselves to be identical turned out to be fraternal when blood tests were used. Even worse, there have been persistent accusations of fraud in such work. All this means that most of the older research on twins reared apart has been discarded. Even so, new work does show that some traits of personality – aggression, introversion and so on – have a genetic component. This does not, of course, mean that nurture can be disregarded. An intrinsically violent man may be calm until he is given a chance to prove his genotype by joining the army.

A more subtle approach involves a comparison of the degree of similarity of identical twins with that of fraternals. As both kinds of twin are brought up within the same household the extent to which they share an environment is, or so it appears, the same. Any greater resemblance of identical twins to each other must then, it seems, show genes are involved.

This approach could be powerful but has its own problems. Both types of twin are brought up together, but identicals may copy each other's behaviour. That makes them appear similar for reasons unconnected with biology. The fact of being identical twins – perhaps with similar names and dressed in identical clothes – may predispose to mental disease.

One of the fundamental assumptions of twin studies – that identicals and non-identicals differ only in the extent to which they share genes – is not always justified. Life before birth can be tough; and more so for identical twins than anyone else. Many illnesses of adult onset – heart disease being one – are, at least in part, due to difficult conditions in the course of the pregnancy. Twin pregnancies are always more of a strain than those of a single foetus. As a result, a shared and hostile environment may impose more similarity on a pair of twins than expected on genetic grounds.

Identical twins come in two forms. All arise from the splitting of an early embryo. Some are mirror-images of their sib; to look at one is to see the other in a mirror. Such individuals divided quite late in development, when the left-right pattern of the embryo had been set down. A late split increases the chance that the twins will share a common placenta and will have to fight for a share of their mother's blood. Such twins survive much less well than those who have a placenta each and the survivors are born ten days or so earlier than those with a less difficult time before they are born. So intense is their struggle that one may steal blood from the other, so that one grows up large while the other is small and anaemic. To complicate matters further, some non-identical twins – no more similar in their genes than are brothers or sisters – also exchange cells early in development, so that each is a chi-maera, made up in part of their sibling's tissue.

Nevertheless, to compare the two kinds of twin has had its successes. Members of a pair of identical twins are twice as likely to suffer from coronary heart disease than are those of a pair of fraternals. For diabetes, the figure is five times. Even tuberculosis is shared to a greater extent between identical than fraternal twins as a hint of an inherited susceptibility. Other characters, such the age when a baby first sits up, are also more similar for identical twins.

The argument about nature and nurture is of more than scientific interest. It has been rehearsed endlessly by those with one or other political axe to grind. Genetics once used an axe sharpened in the fires of Social Darwinism. Now that it has hit the headlines, there is a new acceptance of biological theories of human behaviour. Arson, traditional-ism and even zest for life have all been blamed on the DNA. The nineteen-sixties were the decade of caring and a child's inability to concentrate was blamed on poor teachers. Then there was the 'working-mother syndrome' in which a parent's absence was held to be at fault. Now some psychologists have invented a whole series of behavioural ills coded in the genes while others place renewed weight on friends (and not parents) as the main agent of a child's development.

Psychology's obsessive need to dissect biology from experience is alive, well and as simplistic as it ever was. One study finds that students with hay fever are unusually shy. This proves that 'there is a small group of people who inherit a set of genes that predispose them to hay fever and shyness'. That is naïve; but family and adoption studies do suggest that some aspects of personality, from introver-sion to the speed of response to a sound has an inherited component. From there to the discovery of any genes involved is a long step; but psychologists have not been shy about taking it. Announcements of the discovery of single genes for manic depression, schizophrenia and alcoholism have all quietly been withdrawn.

One form of behaviour has always raised passions about gene and environment but makes an excellent case that genetics and social attitudes have little to do with each other. Homosexual attraction is almost universal at some time in every lifetime. Some people continue to prefer their own sex. Exclusive homosexuality is a convenient subject of study for those interested in the genetics of human conduct as it is easy to identify, quite common, and no longer much concealed.

One study of American male homosexuals hinted at an association between such behaviour and a group of genes near the tip of the X chromosome. First, the brothers of gay men were more likely to be gay than are males in the general population. This does not in itself say much, as brothers share an environment as well as genes. However, gay men's relatives on the mother's side were more liable to be gay than were those on the father's, suggesting that the trait is passed through females. Again, this is not proof of an innate predisposition (even if it implies a possible gene on the X chromosome). The best evidence seemed to come from the X chromosomes of pairs of homosexual brothers. Most who took part in the study shared a particular segment of DNA towards one end of that chromosome. Somewhere in its hundred or more genes may, it was suggested, be one that inclines some carriers to that form of sexual behaviour.

After an initial burst of publicity, the result proved hard to replicate (as is often the case for such characters, in which different genes might be involved in different families) and the simple idea of a 'gay gene' is now dismissed. Whatever the science, the main interest lay in the response by some – but not all – of the gay community. Many, it transpired, were happy to use biology as a justification for their way of life. The idea that sexual preference was inherited meant, they concluded, that sexuality was not contagious and that battles by bigots to dismiss homosexual teachers were not justified. More important, it gave a welcome sense of separation: of a shared difference that was present for reasons beyond individual control. All this disconcerted the many biologists who had spent years fighting the idea that sexual preference, crime or poverty are inborn and cannot be altered by social means.

This new hereditarian orthodoxy, like the old liberalism, asks too much of biology. It echoes a forgotten dispute of the 1930s. The German geneticist Theobald Lang claimed to have found that the sisters of gay men had somewhat masculine characters, and that male homosexuality might therefore be inherited. Whatever the accuracy of his claim that hint of a 'gay gene' gave rise to two quite opposed (albeit equally logical) responses. The Nazis, needless to say, took the brutal view: 'they are not poor sick people to be treated; they are enemies of the state to be eliminated!' In contrast – and faced with the same information – the socialist medical association (then in exile) wrote that 'Homosexuality is inborn and not subject to the free will of the individuals who come into the world with this inversion. The laws against it should be abolished.'

Like some members of today's gay community, both left and right felt that if that behaviour was innate it must be outside the control of those who display it. Each political group saw its response – eugenic sterilisation versus liberal legalisation – as consistent. Neither asked what is meant by a gene 'for' something, homosexuality included. The story of the German 'gay gene' points up the irrelevance of genetics to political opinion. Whatever inherited basis a character may have, preconceived views about its merits will not be changed by science.

Nowhere is the difficulty of separating science from politics, and the confusions of nature and nurture, more malign than in the study of differences among human groups. Older textbooks on race sooner or later come to the question, always treated with a certain prurience, of inherited differences in intelligence. That such differences existed and that they were inborn once seemed obvious. Linnaeus himself classified humans as *Homo sapiens*, thinking man. For the species as a whole, he could be no more precise in his definition than *Homo, nosce te ipsum*: Man, know thyself. His description of the different varieties of humankind, in contrast, used behaviour as a character. Linnaeus' definition of an Asian, for example, was someone who was yellow, melancholic and flexible. Even forty years ago, racial stereotypes of the most predictable kind were still the norm.

Much of the work on inherited differences in intellect among races is contemptible and most of the rest is wrong. The wrong argument goes like this. Blacks do less well than whites on IQ tests, so that they are less intelligent. The IQ scores of parents and children are similar, so that differences in intelligence are controlled by genes. The difference between blacks and whites must therefore be set in DNA.

This argument is deceptively simple. It was once used in the USA as an excuse not to spend money on black education, and a variant of the theory, which sees poor rather than black children as victims of their genes, is employed in Britain by those who resent investment in state education (although, oddly enough, the most devoted hereditarians improve their children's environment by sending them to private schools). Simple as it may be, the argument is utterly false.

Whether IQ tests are an unbiased measure of intelligence is a matter for those who design them. The general consistency in the ability to move shapes around, or to do simple language puzzles and sums suggests a certain objectivity in the measure. The similarity of parents and children in their ability to do the test does not in itself say much, as families share the same environment as well as the same genes. It would be surprising if there were no genetic component in IQ variation. Many with low IQ suffer for genetic reasons, as several inborn illnesses manifest part of their effect by damaging the brain. Although normal variation in intelligence may not be related to such genes – after all, inborn blindness has nothing to do with variation in colour perception – a few inherited diseases do alter specific parts of the IQ mix. One, Williams syndrome (which involves the loss of a tiny section of chromosome), causes heart problems and a rather odd appearance; and a complete inability to deal with objects in space. Patients asked to draw a bicycle do a reasonable job with the wheels, the handlebars and the pedals – but they are scattered on the page. They find it impossible to arrange the parts into an image of the whole machine. However, their ability to speak or to do sums is not much affected.

This rare disease suggests that separate genes affect different parts of the IQ mix but, as always when using the abnormal to study the normal, says little about variation in the population as a whole. A mass of evidence from twins and adoption does suggest an inherited component to IQ. Indeed, a variant form of one gene involved in the growth of cells is frequent among children of very high intelligence (although it explains just a small part of the total variation). The gene involved helps move enzymes around inside cells and those of high intelligence may be more effective at burning sugar in the brain. Some claim that as much as seventy per cent of the variation in score within a population is due to diversity in its genes. This figure seems high, but can be accepted for the present. At first sight it looks like powerful evidence for the view that any racial differences in IQ must be set by biology.

In fact it has no relevance to understanding whether such differences – if they exist – are inborn or acquired. Why this is so can be seen in another character. In the United States, the blood pressure of middle-aged black men is about fifteen points higher than that of whites. Twin and other studies show that about half the variation in blood pressure within a group is due to genetic variation, and some genes that influence the character have been tracked down. The figures for blood pressure look similar to those for IQ although in this case blacks come out with a higher score.

Doctors and educationists have a subtle difference in world-view when faced with such figures. Doctors are optimists. They concentrate on the environment, the fact that blacks smoke more and have poorer diets than do whites, and try to change it. In the USA, optimism has paid off and high blood pressure among blacks is less of a problem than before. Many educationists are less hopeful. To them, the existence of inherited variation in intelligence removes the point of trying to improve matters with changes in the environment. Blacks, they say, have worse genes. These cannot be altered, so that it is futile to spend money on better schools. Their theory has been proved wrong. Over the past thirty years the average IQ of Japanese children has risen to ten points higher than that of Americans. Not even the most radical hereditarian claims that this is due to a sudden burst of evolution in the Far East. Instead, the schools are getting better.

Both genetical and environmentalist views of blood pressure or IQ are naïve. Characters like these are shaped by both gene and environment and it is meaningless to ask about genetic differences except in populations that live in the same conditions. I once did a simple experiment with a group of students. I divided them on the basis of hair colour. The fair-haired group were sent downstairs for coffee. The other set measured their own resting blood pressure. I then summoned the coffee-drinkers. As they had just run upstairs and were dosed with caffeine, their average score was higher than that of the dark-haired students. There was an association between blood pressure and hair colour.

Family studies show that much of the variation in resting blood pressure is due to inherited variation. To most of the students this proved the existence of a genetic difference in blood-pressure between dark fair-haired people. Only when let into the simple secret of the differences in the experiences of each group was it obvious what is wrong in that claim. The students had made the same mistake as the educationists. High heritability combined with a difference between groups need not say anything about genes. The race and IQ story is, in the main, one of a dismal failure to understand basic biology.

A belief in heredity, rather like faith in predestination, is a good excuse for doing nothing. At least the environmentalist version can be used to try to improve matters. The genetical view is often taken as a chance to blame the victim; to excuse injustice because it is determined by nature. In the last chapter of *Daniel Deronda*, biology wins. The hero returns to his ancestral roots and marries Mirah Mordecai, with the Cohen family in attendance. His admirer Gwendolen Harleth is left to console herself with the memory of her unlikeable mate Henleigh Grandcourt, drowned a few pages earlier. Determinism triumphs, which is convenient for the novelist. Fortunately, real life is more complicated than that. One of the most remarkable discoveries made by the new genetics has been to show how little we understand about the human condition that we did not know before.

Chapter Six

BEHIND THE SCREEN

If, by chance, you are reading this book on a train, or in a library, or anywhere apart from in solitude, steal a glance at the person to your left and to your right. Then, comfort yourself with the knowledge that two of the three of you will die as a result of errors in (or side-effects of) the genes you carry. Should that idea be unwelcome, it is worth remembering that a century ago, in that train or library (and depending on the age of your companions) two of the three of you would be dead already.

Life, towards the end of the second millennium, underwent a great change. A British baby born today who lives through the difficult first six months has only about one chance in a hundred of failing to make it to adulthood. In Victorian times – and long before – the figure was, for most newborns, about one in two. In those days, death came from outside: from starvation, infection, or cold. The eugenicists were concerned about the inborn weaknesses of future generations, but in fact most of their fellow citizens died for reasons not directly connected with their genes. Now things are different: we have won the battle against the external world and face the enemy within; our innate failings, central as they are to ailments such as heart disease, diabetes or cancer. As a result, most of us nowadays die of a genetic disease (although not many notice). Genes impinge upon us more than they did in Galton's timeand our new ability to read their message may alter lives and deaths in unexpected ways.

Both Galton's Laboratory for National Eugenics and Davenport's Eugenics Records Office (which changed its identity through a merger with the Cold Spring Harbor Laboratory) are now world centres for human genetics. They and the hundreds of research groups that descend from them have come up with the technology for searching out genetic imperfection that Galton and Davenport lacked. Many of the questions that obsessed the biologists of a century ago have been answered. What is the relationship between people and genes now that we may soon have the tools to identify the inadequate and to carry out some kind of eugenical programme if one was called for? Will we screen all babies at birth; or is that a step too far?

No serious scientist has any interest in a genetically planned society. But the explosion in genetics means that, like it or not, we must face ethical problems of the kind so comprehensively ignored by its founders. Can, or should, choices be made on the basis of DNA? What is the balance between the rights of individuals and of society in the light of the new biology; and is there any need to worry – as the eugenicists thought – about those of unborn generations? One intellectual hero, Sam Goldwyn, dismissed the issue by asking 'what did posterity ever do for me?', but his predecessor, Plato, saw a moral duty to the future in that 'mankind gains its hope of immortality by having children'. To interfere with the genes of the present has an effect on that future and to diagnose an error in one individual at once draws in his family; those alive and those yet to be. Where should the duties of science end?

Genetics has undergone a healthy shift in attitude. Most of its practitioners are not concerned with the quality of the distant future. They feel responsible to people rather than to populations; to today and not to tomorrow. Biologists are, indeed, more cautious about their work than is the public. In one poll three out of four Americans found the idea of inserting genes into human sperm or egg quite acceptable but almost no scientist would contemplate the idea.

In the nineteenth century the bacterial theory revolutionized medicine. Some hope that DNA will do the same in the twenty-first. Genetics might help to predict disease before symptoms appear, prevent it before damage is done or even cure it by molecular microsurgery. Whether or not it succeeds, it will reveal many secrets. DNA may have shattered Plato's notion of the perfect human, but reminds us of his notion that men may be classified by their very nature – not just into those of gold, silver, iron and brass; but into thousands of classes, each at risk of certain diseases, of certain environments, and each, perhaps, endowed with some unique and inborn talent. Are we ready to expose the skeletons hidden in every genetical cupboard? Mass screens for genetic defects are in the air (with the British government among the first to offer its population for the task). The eugenicists would have been happy with the idea and libertarians are alarmed; but now it seems that the job may be more difficult than anyone had hoped – or feared.

For much of the time genetics deals with healthy people, either carriers of single copies of recessive genes, or those with damaged DNA that might affect their future health. By so doing, it draws more and more under the aegis of medicine. Genetics was once a science of the exceptions. Dreadful as inherited disease might be for the families involved, it did not seem to impose upon most people. Genes are responsible for severe inborn defects, but most are impossible to treat (so that those affected die young) and each is rare. With an overall incidence of one or two in a hundred live births, genetic problems seem a minor part of the history of death; crucial to a few, but irrelevant to the many.

For such rare diseases tests do pay, to use that term in its crudest sense, because of the effectiveness of pre-natal diagnosis and pregnancy termination. In Holland the national counselling service costs about thirty million pounds a year. It prevents the birth of from eight hundred to twice that number of severely handicapped children. Even in that small and efficient system, the expense of their lifetime healthcare would be between about a quarter and three-quarters of a billion pounds. For Britain, with a population four times as great, the figures must be multiplied in proportion. In the United States, for fragile X syndrome (a common cause of inborn mental defect) the cost of prevention of a single birth is $12, 000 compared to the million-dollar cost of support.

Such calculations sound offensive, or even brutal, but equations like these are commonplace in medicine. To balance cash against quality and length of life is unavoidable; and the sums may be stark. Even so, given that so many single-gene conditions cannot be treated and that many of the others demand permanent care, they are relatively simple to cost. At first sight, and forgetting any moral dimension, the equations seem clear, both for pre-natal diagnosis and for the care of those born with an inherited illness. For genetics, though, costs and benefits have ambiguities of their own. Now that DNA has entered the domain of common diseases it will allow early diagnosis of conditions, treatable and not, that come on in later years. As it does so it may produce a whole new social class, the healthy ill, who – hale and hearty as they are until the fate coded into their genes makes its presence felt – turn to doctors for help that they cannot give.

The biggest difficulty may prove to be diagnosis by proxy, the inadvertent discovery that a third party, a relative, is at risk. Should doctors inform other family members, even those outside their own care, of their situation? Already physicians have been sued for not telling relatives of a death from inherited colon cancer because the information might have allowed them to protect themselves. In the USA some states see doctors as responsible for informing the wife of a psychopath of her own risk and insist that physicians must tell the rest of the family about inherited disease. Others feel that to inform the patient is enough and leave it to him what he does (the practice recommended by a British Select Committee). Most people, when asked, agree that relatives should be told about inborn conditions and this may become common practice – which changes the normal laws of confidentiality. And how long must a hospital keep in contact with a patient? As more accurate tests emerge, as they will, those who once scored negative on a DNA checkup might then not do so.

Genetics hence calls for decisions about what information should be gathered, by whom, and to whom it should be available. It shifts the boundary between private and public responsibility. To what extent do duties extend within a family? Where do medical decisions end and the values of society take over? Fortunately, perhaps, the more we learn, the more unlikely the notion of a universal screen for imperfection appears.

Without doubt, genetic tests may be helpful. To terminate a pregnancy can be a great relief to a woman found to be carrying a damaged foetus. On a more positive note, any parent with a child with phenylketonuria or inherited colon cancer (each of which can be treated) knows how important genetic checks can be. Perhaps screens of whole populations will find many of tomorrow's patients. They could identify those at particular risk, be they for smoking, stress, chemicals, or certain foods. The new insight into how DNA interacts with the world outside may at least change attitudes to risk. Most people know that to smoke causes cancer and that a fatty diet may lead to heart disease. Certain genes predispose their carriers to the harmful effects of tobacco or fat and some individuals may be able to drink, smoke or eat lard with impunity. Propaganda about smoking and lung cancer has not been very effective. Those exposed to it have an impressive capacity to assume that if one smoker in ten contracts the disease, then that will be someone else. If it is possible to identify exactly who will get cancer if they smoke, individual terror may prove a better deterrent than collective risk. One form of the gene for a protein called alpha-1 antrypsin much increases the risk of emphysema among smokers. The incidence of those with two copies is around one in five thousand; and half of all smokers unlucky enough to fall into that group die younger than forty, whereas the lifespan for gene carriers who do not smoke is extended by twenty years. To discover one's individual danger concentrates the mind: and in Sweden one in ten teenagers with the high-risk genotype smokes, compares to a fifth of others.

To assess future health is not new. All doctors check their patients for high blood pressure and healthy women are examined for signs of breast cancer. As technology advances, signs of other cancers (such as prostate cancer, which generates antigens in the urine) can be checked long before symptoms appear. DNA will make diagnosis easier, with tests for a wide variety of diseases, either inherited or the result of mutations in body cells themselves. Once a patient or someone worried about a family problem turns up, DNA is just another weapon in the doctor's armoury. Medicine has already opened up, with home blood-pressure and cholesterol kits. Home pregnancy tests are in some senses already genetic screens, for many positive results end in termination, whatever the genes of the foetus. In the United States, over-the-counter tests are available for cystic fibrosis carrier status and for genes that predispose to breast cancer.

The technology to sell many more has arrived. Gene chips, as they are known, set out an array of probes for thousands of different genes at once. The idea began in the 1970s. A particular sequence of DNA is held in a stable matrix, and possible matches, each one labelled with a radioactive probe or a fluorescent dye, floated past until one binds to it, in a sort of fishing for genes. Now, ten thousand bait sequences can be put on a single glass slide using methods developed by the computer industry. In principle one chip could test for all the common genetic conditions at once; and, although no doubt they will be expensive, such devices may soon be on the market. The companies involved, some say, exaggerate risks and play on fears to increase sales. A test for a hereditary breast cancer gene costs $2500, far more than it costs to produce. Even so, their use will be hard to control, and in surveys most people think that they should be available. The interval between the discovery of a gene and the sale of a test is short and doctors will have to deal more and more with those who have, rightly or wrongly, diagnosed themselves or their children as at risk.

Medicine might gain from such information as it allows treatment to be targeted more accurately. To tailor pill to patient may become common. The ability to deal with anti-cancer drugs varies fifty-fold and a dose helpful to one individual may be fatal to another. For a certain drug used against leukaemia one child in ten has a low tolerance, so that the screen is essential. On the other hand, drugs not much used now because of their toxicity may turn out to be safe for some. The new approach might also change people's jobs. Nobody wants to be a passenger in an aeroplane with a colour-blind pilot and those with the gene do not get the job, for their own safety and that of others. That logic could extend a long way. People who inherit certain forms of the alpha-1 antitrypsin enzyme find it hard to deal with dust or pollution. In the same way, some individuals with particular forms of the proteins used to make poisons safe are more susceptible to industrial carcinogens. Genetic screens might become part of employment and the tests become a company's duties. However, genes might also be used as an excuse not to improve the environment. The journal *Chemical Week* once wrote that '... it makes no economic sense to spend millions of dollars to tighten up a process which is dangerous only for a tiny fraction of employees ... if the susceptible individuals can identified and isolated from it.'

In the United States and elsewhere, employers pay for their employee's health insurance. Insurance, of any kind, is a mechanism for diffusing risk. The cost of an accident is diluted by sharing it with those who never make a claim. Those who enjoy driving drunk or keeping gold bars under the bed pay more and do not complain (much) that their lifestyle forces them to do so. But what about health insurance? In the USA (and more and more in Britain), access to medical care is limited by the ability to pay. Twenty million Americans must buy their own cover and tens of millions have no health insurance at all. Anyone buying a policy is asked to disclose any medical problems of which they are aware. Many applications are denied, and for the remainder all 'pre-existing conditions' – reported or not – are excluded.

Genetics raises interesting questions. Should an insurer have the right to demand the results of a test to help them decide the price or whether to offer cover at all? Is a damaged gene a "pre-existing condition"? After all, everyone must die; and DNA does no more than tell some (and, in years to come, perhaps most) people when that might happen. But, as insurance depends on spreading risk, genetics may be a terminal blow. It erodes our ignorance of the future. No-one will play with a gambler who knows his opponent's cards and no-one will pay for cover when they are certain that they will live to a ripe old age (and will not need it). The same may be true when the insurers know that an expensive illness is programmed into the genes. Insurance already suffers because people at high risk are more likely to buy a policy. Genetics might spark off a war of cost escalation that ends with only the risk-prone paying for medical insurance.

Already, the companies refuse to cover those doomed to Huntington's Disease and other ailments. Denying cover is no empty threat. A woman in charge of a fragile X screening programme in the United States was refused insurance because her children had symptoms although she had none. In another case, the insurer agreed to pay for a foetus to be tested for cystic fibrosis – but only if the parents agreed to have an abortion if the test was positive.

A commercial health market sees good and bad buys and an employer who pays the bill faces pressure not to hire someone who may be in jeopardy. All this is an argument for a national service which diffuses risk among the whole population. Then health care will revert to the role of a policeman rather than a security-guard, with an acceptance that all must pay, even if some are in more danger than others.

Public health, like insurance, also involves a balance between individual rights and social obligations. Why, after all, is it no longer acceptable to spit in the street; and why do smokers regard themselves as a persecuted minority? Perhaps to screen a new-born child, or an adult, for its inborn weaknesses will come to be seen as a duty both to the person involved and to the social order rather than a mere commercial transaction. Already, every infant, with or without the knowledge of its parents, is checked for whether it carries a range of genetic diseases, PKU being only one.

There are plenty of cases in which this could be useful. Hemochromatosis is a recessive condition that leads to a failure to cope with iron in the food. Untreated, it may be fatal, with damage to heart and liver. About one European in four hundred is born with the disease, and it is even commoner among those of African ancestry. Hundreds of thousands of Britons are in danger (with men at five times the risk of women). The gene has been found and most patients carry one of two mutations. Treatment is cheap and simple. A vessel is opened to lose blood and to prevent the build-up of iron; medical wits have it that patients should grow roses as blood is such a good fertiliser. A simple scan early in life could save many lives and save much expense involved in the treatment of those to whom the damage has already been done.

Medicine has also succeeded in treating genetic diseases that once killed young. The main concern of the eugenics movement was the biological future. Sterilisation was an easy way to reverse what they saw to be undesirable trends. The new biology hopes to become a positive force rather than a mere filter for the imperfect. Cystic fibrosis is lethal because the lungs fill with mucus, and because certain digestive enzymes cannot be made. Conventional medicine – careful treatment of the lung problem, and the use of an enzyme that helps break down mucus – has increased both the quality and the length of life of those with the disease. For some patients a heart-lung transplant can help.

Such successes mean that medicine has already altered the genes of years to come, for many of those who once would have died of cystic fibrosis and other diseases can now be saved to pass on their faulty DNA. Opticians, too, have played a part. A short-sighted hunter-gatherer may well have starved, but the invention of spectacles removed the penalty attached to any genes involved and they have gained as a result. Only if spectacles were banned would that cause problems. Any success in treating inborn disease causes the gene to become commoner in later generations; but as long as medical treatments remain available that will have little practical impact.

Whatever these local triumphs, the biggest problem of modern genetics is one that the public has scarcely realised; the ubiquity of inborn disease. About one child in thirty in Britain is born with an overt genetic problem of some kind and inborn illness causes about a fifth of infant deaths. Over a third of blind people face their plight for genetic reasons and more and more illnesses have been revealed to have an inherited component. In some places the problem is even greater. Around the Mediterranean and in Africa errors in red blood cells, which evolved to protect against malaria, affect millions. Cyprus has many genes for various forms of thalassaemia, loss of a segment of the haemoglobin molecule. Any child born with two copies suffers from severe anaemia. The treatment is blood transfusion, which works but is so expensive that to treat all affected children might soon soak up half the health budget. One person in fifteen, worldwide, is a carrier of one of the malaria resistance genes. Without a medical breakthrough no society will be able to treat the millions of anaemic children who will be born unless something is done to reduce their number. High cost will mean hard choices.

Most genetic technology is simple: to identify a damaged gene and offer the choice of therapeutic abortion. All the common defects can be detected in this way and many more have the prospect of a test. But, as the screens become more sophisticated, where should the line be drawn? In Russia pregnancies have been terminated because the foetus is thought to carry genes that dispose to diabetes. But diabetes is a disease that can sometimes be treated with insulin. And what about diseases for which no treatment is yet available, but might be curable by the time the child is in danger of dying? In muscular dystrophy, for example, we now know what protein has gone wrong and a normal version can help mice with a gene for a similar condition. It is not impossible that some treatment may be available within the next couple of decades. As many boys born now with the disease are likely to live as long this poses a moral dilemma of its own. For pre-natal diagnosis, the equation is affected by the age of parents (and hence the mutation rate), how related they are, any family history of inherited disease, the severity of symptoms, the possibility of treatment, and attitudes to abortion. All this make the process more and more ambiguous.

Decisions based on appraisals of inborn quality are not new but now, for the first time, may be accurate. Should damaged genes be allowed to pass to the next generation, or should the human race attempt to enhance its quality in some way? There are each year, worldwide, about ninety million births and sixty million induced abortions. Britain alone in 1998 had a hundred and eighty thousand abortions, only two thousand of which were on grounds of abnormality. Many more pregnancies end without the woman knowing of her condition, often because the foetus has a severe genetic defect. Even more eggs are lost. A baby girl has a million or so in her ovaries. Three quarters disappear before puberty, and at the age of twenty five she loses, on average, forty or so a day while producing only one a month. The waste of sperm is even more prodigious and, for either sex, many of the sex cells are doomed because of their biological weakness. Genetic selection is a natural part of reproduction. Even so, attempts to choose sperm or eggs or to change the balance between uncompleted and completed pregnancies lead to bitter controversy. Some demand that the state control reproductive choices but others feel that such decisions must be the parents' alone.

Galton, no doubt, would approve the many conceptions that are ended for genetical reasons. In the end, what genetic screening achieves will be limited by attitudes more than technology. They can be hard to predict. In the bad old days, Germany was much at fault; but liberal Sweden sterilised sixty thousand in a programme that lasted long after the end of the War. As late as 1995, South Australia's Reproductive Technology Act forbade treatment for infertility to those who had faced criminal charges. Britain, in contrast, the home of the whole idea, never put eugenics into practice. Ninety-five per cent of the twenty-first century's children will be born in the third world. Although in most places genetics has still to make an appearance, China already has a well-established service. In 1993 the country passed a law that was designed to 'put a stop to the prevalence of abnormal births and heighten the standards of the whole population'. Most of the country's geneticists are in favour of compulsory tests before marriage, and of pre-natal diagnosis of severe genetic diseases (with the implication that a termination is called for).

Elsewhere, though, attitudes can be unexpected. Sardinia is a rather traditional Catholic society in which many marriages risk having a child with thalassaemia. Nine tenths of the couples who face that predicament now know this; and when the woman has an affected foetus nine tenths of those choose termination. Tests offered to older mothers in Denmark led to a fivefold decrease in the number of Down's Syndrome children. Perhaps illnesses such as Huntington's will soon become rare as those at risk decide not to reproduce. Some of the most enthusiastic campaigners for tests for inborn disease are parents who have had an affected child. That in itself says something about where the ethical balance lies. In some places though, such is the passion for the 'right to life' that charities who appeal for funds for genetical research never mention the idea of abortion, but instead concentrate on the (often hopeless) search for a cure.

Genetics as a negative force embodies another subtle tyranny; the dictatorship of the normal, the pressure to produce an average child. The United States has seen demands for growth hormone to be given to children who grow up a few inches shorter than average and would once have been accepted as ordinary. One in ten Britons would consider termination of pregnancy for a foetus found to have two missing fingers; and, in a mirror image of genetic discrimination, a majority of deaf people claim that they should have the right to prefer the birth of a deaf child who might fit better into their family. Achondroplasia, the common form of dwarfism, arises from a dominant mutation. It has some effect on the health of those who bear the gene, but most of the time they are well, and many are positively proud of their condition. Almost all cases arise from new mutations and are born to parents who had no idea of the risk. The gene has been found. Any pregnancy in which the child appears to be growing slowly is monitored. At first, to find that the growth problem was due only to achondroplasia was seen by doctors as useful reassurance to the parents. Many physicians were alarmed to find that the response was often to demand a termination. Now, some centres do not reveal the results – but who has the right to conceal the truth?

Achondroplasia is a reminder that genes involve people as much as DNA. That lesson has been learned again and again. The first attempts to use the new knowledge ran into difficulties because they ignored social realities. A search for carriers of the sickle-cell gene in the American black community thirty years ago led to great bitterness. Although carriers are quite healthy except in conditions of extreme oxygen shortage (which few experience) some states made screening compulsory. Those carrying one

of the mutation were discriminated against in jobs
for insurance. Even worse, people who did not carry
sickle cell considered those who did to be less healthy and
less happy than did the carriers themselves. The carriers
were discriminated against when it came to marriage and
the programme gave hints of eugenic attempts to improve
the quality of the population rather than the health of
individuals. The scheme – albeit conceived with the best
of motives – was a model of the way in which genetic
information should not be used.

Even so, the genetic inquisition is here to stay. Why not,
after all, extend testing to the adult population, for their
own good or for that of society and leave society to work
out how to deal with the difficulties as they arise? The
history of discrimination against the genetically unfortunate is a miserable one: but, after all, that was long ago.
Would not someone at great risk of heart disease like to
know before real damage is done, in time to choose his
vices to reduce the risks or to ask for early treatment? The
idea is seductive and much-discussed. The truth, alas, is
not always so simple, for several reasons.

How far testing can go is set first by the pervasiveness
of imperfection. Any recessive disorder always involves far
more carriers of a single copy of the variant than of two.
If an illness of this kind affects one birth in ten thousand,
about one normal person in fifty carries the gene – which
means hundred times as many copies in healthy people as
in those who are ill. As a result, the notion that one can
improve the long-term health of the population by preventing those with inherited disease from having a child is
simply wrong. Almost everyone carries one or more different genes for recessive inherited disease. A universal screen
would provide information which is unwelcome and is of
almost no use.

Take cystic fibrosis. One British child in two thousand

five hundred is born with the condition, so that about two million Britons carry a single copy and in a tenth of marriages one or other of the partners is a carrier. Any screening programme would turn up millions of such couples. If married couples were checked for all recessives, so many carriers would be detected that it is hard to know what to do with the information – or whether it was worth gathering in the first place.

That problem is compounded by a dawning realisation that screening itself may be much more difficult than was once hoped. Mendel's laws are so straightforward that the map of the genes should, it seems, make it easy to pick out those at risk. The public – and many doctors – believes as much and demand is strong. It seems that a new era of certainty is near, but at least where genetic screening is concerned the truth is more ambiguous.

There are two main difficulties in this application of simple rules to complex problems. First, for most genetic conditions, the DNA involved can be damaged in a number of ways. Every population – sometimes, every family – may have its own mutation in different parts of a structure that may be tens of thousands of bases long. A test that detects an error in one group hence may not work for others. As a result, instead of a universal screen, many separate checks may be needed. To speak of 'the' test for – say – cystic fibrosis means less than it seems. More important, many inborn conditions, although they run in families, involve several or many genes that come together each generation in shifting constellations whose effects are hard to predict.

For single gene conditions such as cystic fibrosis some mutations happened long ago and have spread to millions of people. Others took place recently, and are found in just a few. In general, the older a mutation the wider it spreads and the easier it is to generate a useful test. For the hundred or so Mendelian diseases so far studied in detail, the news is not particularly good: most have high diversity, most individual mutations are rare, and – at a guess – the majority of errors have appeared within the past two thousand years and are still restricted in their spread.

All this means that to establish whether someone is a carrier of a simple error, without prior knowledge of the mutation involved, may not be easy. Ironically enough, in alkaptonuria – the first inborn error to be discovered, with its simple pattern of inheritance – almost everyone who bears this rare condition has the same unique mutation. For the more frequent illness cystic fibrosis, however, well over a thousand different mutations are known. Some are quite common. One among them is responsible for seventy per cent of cases in Western Europe. Two thousand miles to the East, that change is found in just a small proportion of patients. Even in Jewish populations in the United States it is involved in only a third of the damage and in some populations (such of those of North Africa) quite a different mistake causes the majority of harm. One illness, an alteration in a single gene, has, it transpires, a multiplicity of causes.

Life is even harder for the many conditions that do not follow Mendelian rules. A few are affected by genes of moderate effect. One form of heart disease is much influenced by variation in a gene involved in the transmission of messages from nerves to muscles, although several other segments of DNA with a weak effect are also involved. Thousands of women die of breast cancer, and for a few, one of two genes known to predispose to the disease are responsible. These are carried by one in eight hundred or so women and represent hundreds of different DNA changes. Removal of the breasts is of some help to those who carry the gene, but this is a drastic move that is not acceptable to many (although small doses of anti-cancer drugs may also help). All this, combined with the rarity of the gene, the fact that nineteen out of every twenty breast-cancer patients do not carry a mutated version but develop the disease for unknown reasons, and the lack of any general test, means that a population screen would not be worthwhile.

In Ashkenaze Jews some damaged versions of this gene are more common because, by chance, most of the population descend from a rather small group of founders. One in fifty or so Ashkenazim carry one of three distinct mutations that predispose to breast cancer. A screen of the whole population might, it seems, pay off, but even here reality is complicated. Because many women with the disease get it for reasons unconnected with these two genes, and because the disease often does not show itself until middle age, the extra burden of risk faced by bearers of the gene is only about one in twenty. All this means that many geneticists feel that screening for an inherited predisposition to breast cancer is not useful even here.

To make life even more difficult for the hapless screener, what seems to be the same disease may arise from errors in quite different genes and, quite often, a condition that has a simple genetic cause in some patients may have almost no inherited component in others. Many illnesses that once appeared to be one have been subdivided. 'Fever' was seen as a unitary disorder with a single treatment and 'cancer' was much the same. That simplistic view changed long before genetics; but genes have speeded up the process.

In spite of the complexity of mutation, nobody doubts that – say – cystic fibrosis is one disease. However, most inherited diseases are not due to simple errors in one or even a few genes: instead, they are (like fever itself) symptoms of a great constellation of failures in the body's machinery. A single gene may be involved in certain cases but not others, or genes of small effect may band together in unpredictable ways. The DNA at fault may differ from population to population, or from family to family; and an inborn problem may not show itself until it is exposed to a particular challenge. As a result, some conditions appear to have an environmental cause in some patients, and a genetic one in others. Such complexity means that to unravel the big killers like heart disease or obesity will not be easy. Even when the job is done, it is not clear how the information will be of much help in screening.

Take, for example, diabetes. Diabetes mellitus affects one in ten people. The cause, a loss of control of blood sugar, seems simple. The illness causes kidney damage, blindness, heart disease, nerve destruction and death and, in the USA, costs $100 billion a year to treat.

It comes in two flavours. One results from a failure of the pancreas, the gland that makes insulin. That problem is quite rare (affecting about one child in a thousand), starts young and can be treated with the missing hormone. The other, non-insulin-dependent diabetes mellitus, is commoner, appears later, and is resistant to insulin treatment. Six million Americans have this illness without knowing it and it is becoming more common. The two forms, similar as they seem, involve separate sets of genes and present medicine with quite different problems. What is more, each condition itself conceals several ailments, some influenced by inheritance, some not.

The insulin-dependent form was once thought to be caused by viruses, by diet, or even by urban living. None of the ideas was sustained. In fact, genes are involved, with the risk to brothers or sisters of patients twenty times higher than that to the general population. Early-onset diabetes is associated with the systems of recognition on the surface of cells. Those with certain combinations of cell-surface cues face a tenfold increase in risk. The illness, it seems, results from an attack by the immune system on its own body.

A single gene explains a third of the variation in the chance of getting the disease. At least twenty others may predispose to it, some involved in the insulin machinery and others at work in unrelated parts of the cell. Identical twins have only a one in three chance of both being affected, so that something other than DNA (perhaps the poor nutrition of their pregnant mother) is also involved. Insulin production varies in a gradual way from person to person with a threshold at which the illness sets in. The level at which this is set is under genetic control and may be modified by some unknown environmental cause.

Because the genetic component of this form of diabetes is, in some patients, simple and quite strong, a DNA test can be used to identify a proportion of the children at risk and to begin treatment before any damage is done. Even so, most patients – those with genes of minor effect, or those exposed to the unknown environmental stress – will not be revealed by a genetic test.

The people of the Pacific island of Nauru have had riches thrust upon them by phosphate mining. Instead of fish and vegetables they eat fat and sugar. Eight out of ten adult Nauruans have non-insulin dependent diabetes and the island has one of the shortest life spans in the world. Perhaps the local genes for susceptibility to sugar were at an advantage in times when starvation was followed by glut. Not until glut was the norm did they become dangerous.

The biological heritage of Nauru is shared by the natives of the New World. Many Mexican-Americans suffer from 'New World Syndrome, they are fat and have difficulty in controlling levels of blood sugar. The risk goes up with the number of Amerindian ancestors, which at first sight makes a good case that genes are paramount. But the disease is almost unknown among American Indians living in their home communities. It affects them only if they change their diet by moving north. Differences among Americans in the incidence of New World Syndrome arise from both nature and nurture.

This second type of diabetes is even more complicated and more recalcitrant to screening than is the childhood form. Genes do play a part, but they separate populations rather than individuals. Among those with Pacific or Native American ancestry, diabetes runs in families, but no single gene accounts for more than a tenth of any individual's susceptibility. Many variants are involved, scattered through the genome, with little clue as to what they do. Some have at best an ambiguous role: a search in America for eleven genes supposed to be involved in the British population turned up only two.

The most notable aspect of non-insulin dependent diabetes is the power of the environment. A change in diet is to blame. Native Americans' or Pacific Islanders' genes are less able to cope with large amounts of fat and starch than are those of Europeans, and the illness follows. DNA is a less effective predictor than is diet. A change in habits would benefit the whole population, whether or not they are at specific risk of the illness.

For diabetes, what seemed to be a single disease is in fact two, or several, or many, each of which might demand different treatments. Some therapies are successful, some less so; and some involve not drugs but a change in habits. Certain patients may be detected before symptoms appear by virtue of the genes they carry, but others will be missed by any screening programme. A few people will develop diabetes whatever their diet, while others may, in spite of an inherited susceptibility, avoid it because of how they live. The genetics of the illness – like that of many others – involves a minority of individuals with a single gene that predisposes to disease, and a larger number who have drawn an unlucky hand of several low-value genetic cards, which can come in many ways. Together they increase the danger but any one is of little use in prediction. For the adult-onset form, to identify those at high risk may concentrate their minds but in the end the genes have little relevance: to ban cheeseburgers would do more than anything medicine can do.

Many (and perhaps most) diseases are rather like this. Susceptibility will be difficult to test for. Variation in one gene that makes a protein important in controlling fat levels in the blood influences the chances of heart disease (although, of course, diet is also important). It also has a great effect on the chances of pre-senile dementia, Alzheimer's disease. For those with two copies of one form of the gene, the disease usually begins before the age of seventy, fifteen years earlier than those inheriting two copies of a more favourable allele. Most patients have no overt predisposition at all; many of those at high risk die for other reasons before the symptoms appear, and footballers, whatever their genes, are in danger of a similar illness because they head the ball. Heart disease is just as complicated. Some families inherit a tendency towards high levels of blood cholesterol, but almost two hundred different combinations of genes can generate the effect. Risk of the disease is affected by how fat a person is (which may itself have a genetic element), blood pressure and insulin levels as well as the genes more directly implicated. As a result, such families are easier to identify with a simple blood-cholesterol test than with the most complicated DNA technology.

All this means that the study of the inheritance of common illnesses is plagued with results that cannot be replicated. Many claims that particular genes are associated with mental diseases such as schizophrenia or depression have not been sustained on further study. In spite of plans for huge (and expensive) sweeps through the genetic undergrowth, with thousands of sick people scanned to see if their DNA is special, the chances of success in the search for the genes involved are small. As a result genetics will have less of a role in the diagnosis and treatment of common disease than is often claimed. What it may do first is to show that what once appeared to be single illnesses – from schizophrenia to obesity – are in fact many, each of which might need a separate treatment. As a result, the prospect of screening whole populations for those at risk is far away.

Even so, the science can bring unwelcome news to some. Many inherited diseases cannot be treated even if they are diagnosed, which at once raises the question as to who would want a test. Half of those who have a parent with Huntington's disease may contract it. Because of the delay in symptoms appearing, those at risk were once left in uncertainty about their fate. The first signs often appear in a patient's thirties or forties as a general restlessness and depression, followed by involuntary movements and ending in paralysis and death, usually within twenty years of diagnosis. As symptoms may not appear until middle age, those in danger are left in an ambiguous position. In Britain a mere one in ten choose to be tested. For them, doubt is preferable to certainty (even if, among those who do test positive, the fear that many would commit suicide has not been realised). In contrast, half those at risk of inherited breast cancer (where drugs may help although their efficacy has not been proved) take up the offer, and eight-tenths accept a test for familial colon cancer, which can in effect be cured by surgery. For those at risk of a genetic illness whether a condition is treatable is central. For a disease about which nothing can be done, most people see little point in a test.

All this is a tribute to common sense. Some decisions are less rational. One in ten of those offered the chance of testing to see if they carry a single copy of a cystic fibrosis gene by post agrees and one in four when an appointment is made; but almost everyone given the chance of an immediate check accepts. All the great killers in the developed world are influenced by genes; and, in principle at least, genetics might be able to tell many people the probable date of their death long before it happens. Why, one might ask, would anyone want to know?

As the blind seer Tiresias put it in *Oedipus Rex*, 'How terrible it is to have wisdom when it does not benefit those who have it!'. Genetics – science in general – produces knowledge; wisdom, how that knowledge is to be used, demands much more and may come only after long and painful experience. Tiresias himself was struck blind for revealing the secrets of the gods. Those of the genes are still in part concealed: and the two of every three people that they will kill can, for the time being, be grateful for that.

Chapter Seven

THE BATTLE OF THE SEXES

Biologists have an adolescent fascination with sex. Like teenagers, they are embarrassed by the subject because of their ignorance. What sex is, why it evolved and how it works were once the biggest unsolved problems in biology. The pastime must be important as it is so expensive. If some creatures can manage with just females, so that every individual produces copies of herself, why do so many bother with males? A female who gave them up might be able to produce twice as many daughters as before and they would carry every one of her genes. Instead, a sexual female wastes time, first in the search for a mate and then in the birth of sons who carry but half of her inheritance. It is still not certain why males exist; and why, if they are unavoidable, nature needs so many. Surely, one or two would be enough to impregnate all the females but, with few exceptions, the ratio of one to the other remains stubbornly equal throughout the living world.

An obsession with sex goes back a long way. The Venus of Galgenberg, an elegant serpentine statuette without the exaggerated breasts and buttocks of later variations on the theme, is thirty thousand years old. Aesthetic interest in the female form goes back even further. A small pebble from an excavation in Israel has been grooved to resemble a woman's body. At eight hundred thousand years old, it is the oldest known work of art.

Curiosity about the meaning of sex is not new. Plato, in the *Symposium*, suggested that there were once three sexes; males, females and androgynes or hermaphrodites. The third sex was split apart by an angry Zeus and doomed to spend eternity forever seeking its partner: 'Zeus moved their privates to the front and made them propagate upon themselves. If, in all these claspings, a man should chance upon a woman, conception would take place and the race would be continued; whereas if man should conjugate with man, he might at least obtain such satisfaction as would allow him to turn his energies to the everyday affairs of life'. This provided Plato with an explanation for the origin of sex and the sex ratio and a neat explanation of the variety of sexual attractions common from ancient Greece to the present day. Two thousand years later the English wit Sydney Smith had the same idea, although his three sexes were men, women and clergymen.

To define sex is easy enough. It makes individuals who contain genes from more than one line of descent, so that inherited information from different ancestors is brought together. In an asexual lineage everyone has one mother, one grandmother, one great-grandmother and so on in an unbroken chain of direct descent from the ur-mother who began the lineage. Sexual organisms are different, because the number of ancestors doubles each generation. Everyone has two parents, four grandparents and so on. Each sperm or egg has half the number of genes present in body cells and in each the genes are scrambled into new arrangements by recombination. When they meet, the novel arrays come together to produce a new and unique individual. Re-shuffling the genetic message is at the heart of sexual reproduction.

The nature of sex is illustrated by two eponymous heroes of British history, King Edward VII (who flourished in the years before the First World War) and the King Edward variety of potato (which has fed the British working class for almost as long). The potato, unlike the royals, reproduces asexually. Every King Edward potato is identical to every other and each has the same set of genes as the hoary ancestor of all potatoes that bear that name. This is convenient for the farmer and the grocer, which is why sex is not encouraged among potatoes. King Edward himself was a different kettle of fish. Half his genes came from his mother, Queen Victoria, and half from his father, Prince Albert. He himself was a new and unique genetic mixture who combined some of the qualities of the two and of an ever-widening pool of more distant ancestors.

It is easier to define sex than to understand it. Men have made many attempts to justify their existence. Mutation may help to explain why life is not entirely female. A harmful change to the DNA in a sexless being will be carried by all her descendants. None can ever get rid of it, destructive though it might be, unless it is reversed by another change in the same gene (which is unlikely). In time, a second error will occur in another gene in the family line. A decay of the genetic message sets in as one generation succeeds another, just like the decay that takes place within an ageing body as its cells divide without benefit of sex.

In a sexual creature, by contrast, the new mutation can be purged as it passes to some descendants but not others. Quite often, one unfortunate will, by chance, inherit several damaged genes, and all are lost at the cost of a single death. Sex also has a more positive effect: as the environment changes some new combinations of genes may be better able to cope with the novel challenge. New mixtures of genes produce successful individuals who have been dealt a favourable hand and others who inherit a less advantageous set. George Bernard Shaw illustrated this in a hackneyed but accurate phrase. When an actress asked if she could bear his child, who might have her body and his brains, Shaw pointed out the risk of having an infant with her brains and his body. Sex reshuffles life's cards: it makes beautiful geniuses who survive and ugly fools who do not. It is a convenient way to bring together the best and purge the worst and to separate the fate of genes from that of those who carry them.

Recombination is a redemption, which, each generation, reverses biological decay. In some ways, it is the key to immortality; a fountain of eternal youth – not for those who indulge in it, but for the genes they carry. Sex speeds up evolution because each generation consists of new combinations of genes, rather than thousands of copies of the same one. Instead of always drawing the same hand in nature's card game (which might be successful in one encounter but will not be so in all), every fertilised egg has a new deal and a new chance to win. The chance may be a small one, but as so many hands are dealt sex becomes a worthwhile, albeit expensive, gamble against a hostile world.

To abandon males can cause problems. The majority of all-female plants can be used for only a few years. They become so loaded with genetic damage that they no longer thrive, or cannot keep up in the evolutionary race with their parasites who in time prevail. Their lineage has become old and tired. Potatoes show the risks of celibacy. The Irish famine happened because the plants used belonged to an old and sexless variety. In the mid-nineteenth century every tuber in European was descended from one or two introductions from the New World made three hundred years earlier. The new crop soon spread throughout Europe. Louis XVI of France, in an astute exploitation of the rustic mind, put guards on the first fields during the day but removed them at night. The peasants, impressed by the apparent value of the crop, were quick to steal examples and to grow them on their own land. In Ireland in 1840 every adult ate several pounds of potatoes a day (in part because their grain was exported to England to pay rent).

Famine struck with great speed and ruinous effect. In 1845, the Irish Freeman's journal wrote 'We regret to have to state that we have had communications from more than one correspondent announcing the fact of what is called 'cholera' in potatoes in Ireland, especially in the North. In one instance the party had been digging potatoes – the finest he had ever seen from a particular field, and a particular ridge in that field until Monday last; and digging in the same ridge on Tuesday he found the tubers all blasted and unfit for the use of man or beast.' In the next five years, a million and a half people starved. Their crop had been attacked by a fungus, the potato blight, which is sexual and has many generations to each one among its hosts. The parasites evolved at a greater rate than could the potato. Nowadays, plants with new sets of genes are tried every few years to counter this. Other asexual crops, such as bananas, have as yet escaped the fate of the Irish potato (although it cannot be indefinitely delayed). The potatoes were forced into an evolutionary dead end from which the sole escape is sex.

Because of the dangers, rather few animals have abandoned that pastime. They include the odd lizard or fish, but none of our close relatives. Even greenflies, which manage without it for most of the time, require a bout once a year or so. With occasional exceptions such as rotifers (tiny denizens of fresh water, among whom no male has ever been found), all-female lineages derive from recent ancestors with a normal sex life, as a hint that chastity is an evolutionary dead end. Just why abstinence is undesirable is still not certain. In spite of the attractions of the mutation theory the frank answer is that, although the reason for the existence of women is obvious enough, there is still plenty of room for argument about the point of being a man.

The perils of abstinence can be seen in men themselves, as they possess the Y, the only chromosome that has abandoned the hobby. When germ cells are formed, all the other chromosomes line up next to each other – chromosome 2 with chromosome 2 or X with X, for example – and indulge in recombination, the sexual orgy of genetic exchange. In a male, the Y does line up with the X, but its embrace of its fellow is less than enthusiastic. Only the tip of the Y exchanges material with its opposite. The rest of the chromosome is held in a kind of biological purdah, safe from the advances of other genes.

Chastity has had terrible effects on the Y. It has long sequences of meaningless DNA letters, many repeated thousands of times. Perhaps this is a hint of what might happen to asexual lineages if abstinence goes on for long enough. Mutations accumulate and cannot be shed and junk DNA may creep in and prove impossible to dislodge. Apart from its narrow role in ensuring the persistence of men, the Y chromosome is a warning of the dangers of continence.

Other parts of the genome have also moved in the same direction. Comparison of the physical map of the genes with the linkage map (based, as it is, on recombination) shows that some parts of our DNA are sexier than others, at least in the sense that more recombination takes place at these well named 'hot-spots'. What is more, females are on this criterion sexier than males, as women show more recombination than do their partners.

Sex itself raises other problems – if sex, why sexes? If to shuffle together the genes of two individuals is such a good thing why has evolution not come up with a scheme which allows everyone to mate with everyone else? As we are limited in our choice of partners to those of a different sex, to have just two of them seems very inefficient. Almost all organisms (with the exception of a few simple creatures that have dozens of genders) exist as just males and females. This means that just half the population is available as a potential mate. If there were three genders, then two thirds of the group might be accessible, and a hundred different reproductive classes could make ninety-nine per cent of our fellows into possible partners. One answer (and it is only one) involves what seems at first sight the antithesis of sex – conflict.

Males are best defined as the sex with small sex cells, sperm; and females as that with large, eggs. Body cells contain DNA not just in the nucleus, but in the cytoplasm which surrounds it. Some is associated with mitochondria, which have genes of their own. Many creatures have yet more in the cytoplasm. It comes from what were once independent beings which now hitch a ride within cells. This DNA (like that in the nucleus) has its own agenda, which is to be copied and passed to the next generation. The cytoplasm is its territory and like a blackbird or a tiger it defends its homeland against invaders. If sperm and egg were the same size (and each had its own population of extraneous DNA) there is a danger of war breaking out on fertilisation. Then, two sets of cytoplasmic genes find themselves in the same space in the fertilised egg. Just like tigers, one set might attack the other until it prevails. This is expensive and could even harm the genes in the nucleus.

The dispute is resolved because one sex – the males – unilaterally gives up. The sex which surrenders passes on none (or very few) of its cytoplasmic genes (which are excluded from the sperm at fertilisation) while the winner, the egg-maker, passes on large numbers. As in most wars, the stable number of opponents is two, and the existence of males and females (rather than dozens of genders) represents a truce in the battle of the sexes.

Biology now understands why sex is there and why it is limited to the tedious dualism of male and female. The technical revolution in genetics has also shown how simple sex is in – and at – conception and what a complicated tangle it later becomes. Existence is, it seems, in its essence female and masculinity just a modification of the feminine experience. The Y chromosome forces the embryo into manhood. If, for some reason, the Y is absent the foetus develops as a female. Some children are born with an extra X chromosome. Their chromosome set is XXY. They are male, but sterile. People with half a dozen X chromosomes and a Y have been found and these too are male, a reminder of the power of this small chromosome to impose its function on the X.

The discovery of a few males with two X chromosomes helped in the search for the gene responsible. They break the rule that to be a male needs a Y. In fact, in these men (most of whom are unaware of their condition) a tiny part of the Y chromosome has been broken off and attached to an X. This is then armed with the information needed to inflict maleness. Because the transferred segment is small the augmented X was useful in tracking down the crucial gene. The gene is found in all male mammals and is similar to another that determines what passes for masculinity in yeast.

The machinery that decides the sex of a fertilised egg may be simple, but the road to adult gender is a complicated one. Sexuality is a flexible thing. In crocodiles, for example, it is determined by the temperature at which the eggs develop, so that females must lay their clutches in a place with a temperature range which allows both males and females to be produced. In certain fish, embarrassment – or social pressure – is important. A shoal of females is guarded by a male. To remove him leads to a period of confusion, until one of the females changes sex and assumes his role.

Once sexuality gets started, great consequences flow

from it. Most of natural history is the scientific study of sex, as the characters which differentiate birds, insects and flowers from each other are, in the main, associated with reproduction. To compare the sex lives of different animals hints at how sex evolved and why animals indulge in one or other reproductive preference. Although humans are in many ways distinct, it might even be possible to learn something about our own habits by looking at those of other species.

Many people have attempted to draw sweeping conclusions about humankind from studies of the private lives of monkeys and apes. It is always dangerous, and usually futile, to try to explain human behaviour in the simple terms used to study animals. Attempts to do so almost all fall into the 'pathetic fallacy', the literary trap which sees emotions mirrored in the weather or the landscape. Occasionally – very occasionally, as in *Wuthering Heights* – this works, but most of the time it ends in bathos. Anthropology has the same problem. It is fatally easy to read into the animal world what we would like to see in our own, to explain the human condition as an inevitable consequence of our biology. Even Charles Darwin, a veritable Brontë among sociobiologists, was at fault. Hidden in his unpublished notebooks is the damning phrase 'Origin of Man now proved – metaphysics must flourish – he who understands baboons will do more towards metaphysics than Locke.'

Metaphysics is one thing, sex another. The Nobel Prize-winner Konrad Lorenz saw us as 'killer apes' anxious to pass on our own genes by murdering the opposition (which may explain his own flirtation with the Nazis), and any decent airport has a row of paperbacks that purport to explain human nature as the remnants of a history as primates with one or other social preference. Until a few years ago the study of sexual behaviour was little more than a set of unconnected anecdotes. It has been transformed by the rebirth of one of the oldest techniques in biology. Comparative anatomy is what convinced Darwin that men and women are related to monkeys and apes. The new science of comparative behaviour hints at how and why their sexual conduct evolved.

Sex is filled with strife, with the very existence of males and females the resolution of a war to pass on cytoplasmic genes. Further conflict arises as males struggle for mates and as males and females disagree about the time and effort needed to raise young. The conflicts among males lead to the evolution of spectacular organs of attack such as antlers. Other traits – such as a baboon's gaudy face – are more subtle statements of male talent and may evolve because they are preferred by the opposite sex.

There is little evidence (in spite of much prurient speculation about beards, breasts and buttocks) that humans have attributes of this kind but, as in most animals, conflict between human males is greater than between females. To be a man is dangerous. At birth there are about 105 males to every 100 females, but this drops to 103 to 100 at the age of sixteen and in their seventies women are twice as abundant as men. Men have more accidents, more infectious diseases and kill each other more often than do the opposite sex. As might be expected, eunuchs and monks live for longer than do those condemned to a normal sex life.

Our close relatives have different life-styles. From a human perspective, chimps are deplorable but gorillas dull. A male chimpanzee copulates hundreds of times with dozens of females each year. The faithful gorilla, on the other hand, has to wait for up to four years for his female to be ready to mate after she has given birth, and even then she is available for just a couple of days each month. That means intense competition among gorilla males for access to females. A successful male may accumulate half a dozen or more, which leaves many gorilla wallflowers out in the cold and anxious to fight for their reproductive rights. Often, these fights are savage, since what is at stake is the male's evolutionary future. Humans are unlike any other primate as they live in large groups as (more or less) faithful pairs. In this, people are more similar to seagulls than to any ape. The closest in behaviour to ourselves is the pygmy chimpanzee. This forms long-lasting pairs within a stable but small group of individuals and has other attributes not unlike our own (such as face-to-face copulation). The average Frenchman or Briton has ten sexual partners in his life and, as in many primates, there is more variation among men in their success than among women. One in a hundred men is responsible for one in six of the females who have sex.

Monkeys and apes show a good general fit between the size difference of the sexes and patterns of mating. In those species with large harems and angry bachelors, males are much bigger than females, because bulk and aggression help in the battle for partners. Gorilla males are twice as large as females, while the chimpanzee's more relaxed society has taken the pressure off sexual hostility and males and females weigh about the same. The argument from anatomy (restricted as it is in a socially complex animal like ourselves) suggests that humans, with men just a little larger than women, have a history of mild polygamy intermediate between that of chimp and gorilla.

Our own behaviour is flexible and often shifts (as in the recent change towards serial monogamy, constancy within a relationship but more than one relationship in a lifetime). There do seem to be some general rules. Strict monogamy is rare and, in most societies, most men have more than one mate during their lives. Polygamy (one male with several wives at once) is far more common than polyandry, the opposite pattern, although this exists in Tibet. In poly-gamous societies as a few men have many wives some must have none.

There are hints of a more salacious past for humankind than that recorded in the modest difference in male and female size. In many mammals, the struggle between males does not stop at copulation. Sperm compete too. Often a female uses the sperm of the male she mated with last, which means that a successful sperm donor must ensure that no other male mates with her until the eggs are ferti-lised. Dogs, for example, stay paired after copulation because the male is guarding the female against intruders. A more subtle way to help one's own sperm is to flood out the contribution of the previous visitor. Different primates show quite a good fit between the size of the testes and the extent of male promiscuity. Chimpanzees, the Lotharios of the primate world, have enormous testes while gorillas, in spite of rumour to the contrary, are far less well endowed. Humans are not too different from chimps in this respect (which may say some startling things about our past). Real enthusiasts for evolutionary explanations point out that men produce more sperm when they return to their partner after a long absence, perhaps to overwhelm any alien sperm that may have intruded. There is also the question – as yet unanswered by science – as to why, in penis size, man stands alone. There are limits to what biology can explain and this may be beyond them.

James Boswell in his London journal (which reveals him to have been no mean performer in his own right) wrote that 'If venereal delight and the power of propagating the species were permitted only to the virtuous, it would make the world very good.' Darwin, too, noticed that sexual selection (as he called it) might do more than improve a male's ability to defeat his ardent competitors. He was much concerned with the evolution of characters with no obvious biological advantage (such as the peacock's tail or the large human penis). The struggle for sex might, Darwin thought, have subtle consequences. If females prefer, for one reason or another, a particular male attribute (such as a bright tail), then males who have it will reproduce more successfully. The tail or its equivalent will become more common in later generations and the showiest males will once again be preferred. In time there may evolve bizarre structures which are so expensive to the unfortunate males that they can evolve no further. Female choice may, Darwin suggested, be as important a part of the sexual equation as is male aggression.

In his book on the subject, *The Descent of Man and Selection in Relation to Sex*, he went further. He suggested that sexual preferences explained why human races looked so different. It was not that they had evolved to fit the place in which they live, but as a consequence of arbitrary choice of a partner. In different places, those looking for a mate may have made different and quite capricious choices. In time, the people of the world diverged: for example, Darwin speculated, those with darker skins might have been seen as more attractive in Africa and those with lighter in Europe. People do tend to marry others who are similar to themselves in intelligence, colour and body build, but there is no evidence that such choices are important in evolution. On average, men – of whatever racial group – do tend to prefer relatively light-coloured females. If sexual selection was important then the blondes will prevail. As they have not, perhaps Darwin was wrong; or, perhaps, the whole issue of sexual choice is so open to social convention that the argument can never be tested.

Men do tend to agree in their estimation of how attractive a particular female might be. Galton made composite photographs in which the pictures of a number of society beauties were printed one on top of the other in the hope of some vision of the ideal woman. His Ms Averages look rather insipid to the modern eye. The job can now be done by computer. For both male and female faces most people find an image made up of several individuals more attractive than one based on a single person, and the more faces used the more appealing it seems. Why there should be this triumph of the typical is not certain (although some suggest that those with extreme faces might also have aberrant – and less desirable – genes). Faces mixed together even out the differences between left and right. A simple experiment with a photocopier shows how two-faced most of us are. Two left cheeks or two rights often look alarmingly different one from the other. As each side of the face is made by the same set of genes, perhaps the greater the asymmetry the feebler the genes. Models (noted for their sexual attractiveness) tend to have symmetrical features and often reveal as much with a full-face smirk into the camera. Again, sexual choice may be involved, although the evidence is weak.

Any discussion of the evolution of sex seems doomed to stray onto such untamed shores of speculation. Males carry eccentric and expensive ornaments, some say, to demonstrate to potential spouses that their genes are good enough to bear the cost. The idea has been used to explain bizarre patterns of human behaviour. Perhaps men take alcohol, tobacco or stronger drugs to demonstrate to women how tough they are, how their constitutions can cope with mistreatment and how they might make excellent fathers as a result. The small tubes found in the tombs of Maya Indians might have been used to give ritual enemas of toxic drugs to the most powerful men, as a guarantee of instant intoxication and a statement of sexual prowess. The habit did not spread.

Conflict between males for the attention of females is obvious, but there are also plenty of chances for disagreement between the sexes. In some animals, the reluctance of females to accept a new mate, persistent though he might be, arises because males invest less in bringing up offspring. It pays them to mate and run; to try and father as many children with as many females as possible. Females need to be more cautious. As it costs so much to produce a child they should choose the male who will be the best father and reject the rest.

The divergence of interest is sometimes obvious. Some males kill a mother's brood by another male with the aim of making her available to themselves. Among the langur monkeys, most of the young die for this reason. Some species even have a form of prenatal cannibalism. Pregnant female horses exposed to a new male reabsorb their foetuses, a behaviour which may have evolved because of the near certainty that if born they will be killed.

Humans reveal the intersexual struggle in less blatant ways. Their battle is an economic rather than a mortal one. If tribal peoples are any guide, societies with private property are more polygamous, as women prefer the better-endowed as mates. When wealth is concentrated into few hands, society becomes more like that of a gorilla, with the richest males monopolising the females. The philo-progenitive (and opulent) Moulay Ismail the Bloodthirsty of Morocco admitted to 888 children. We in the West now seem to be moving towards the chimpanzees, as most men have at least a fair chance of a Ms Right, but in most societies success is still related to wealth. Among the Kip-sigis people of south-west Kenya a wealthy man may have as many as a dozen wives and eighty children. The more land a man has the more wives he obtains and the poorest males leave the community as teenagers and have no children at all. All women, in contrast, tend to have families of about the same size. In Britain, too, men from higher social groups have more partners than do those less well

The Battle of the Sexes

off. An economic conflict between the sexes means that men provide the capital and women choose where to invest.

The battle of the sexes may explain another unusual attribute of human reproduction. Women are the only female primates who do not make it obvious when they are most fertile. Most female primates advertise the two or three days in each cycle when they are able to conceive. Often, this is accompanied by a frenzy of copulation with a series of males. Before modern medicine, most women (and all men) were unaware of when the fertile period was. Women's reproductive coyness might reflect the change in the economic relation of the sexes which came with the origin of society. It could, some suggest, be an attempt to resolve the conflict between male promiscuity and the female's need to ensure the care of her children. By concealing when she is fertile she ensures constant attention from her mate. If he is not sure when she can conceive then he dare not leave her for a new woman in case another male takes advantage of his absence. This is historical speculation with no evidence for or against it – and, as so often with theories of history, several other interpretations are possible.

Males do, needless to say, contribute to the care of their children, but in most societies the sexes differ in their commitment to the next generation. The mother is usually left holding the baby when a relationship breaks up. The difference can be subtle. Many genetic tests can tell parents whether they carry a harmful gene and whether it is wise for them to plan to have children. In a few cases, the test also tells the parents themselves that they are at risk of illness themselves. Huntington's disease is of this kind. Twice as many women as men volunteer for a test, perhaps because their concern for their potential child's future is greater than that for their own peace of mind.

The battle of the sexes is often seen as regrettable but unavoidable, but most people assume that the bonds between mother and child are driven by mutual devotion. To the cold eye of the biologist the transaction between generations is also filled with conflict, with many chances for the two parties to exploit each other. It is in the child's interest to gain as much attention as possible from its mother. The mother's concern is to provide as little as will allow her progeny to survive. If she is too generous to one, the next may suffer.

Such confrontations, deplorable as they seem, are the commonplace of the animal world. There has grown up in biology the comforting supposition that nature is not really red in tooth and claw and that animals rarely do much harm to other members of their own species. The battle for reproductive success shows how wrong this is. Eagles lay several eggs. If food is plentiful all the chicks are fed, but any shortage means that the last to hatch is allowed to starve or is killed by its sibs. Rats, mice and other mammals often eat all their young when food gets short (a habit known as kronism, after the Greek deity Kronos, who devoured his own children).

Any mother is certain that all her children (first, second or third born) carry her own genes, but it is quite possible (and in many animals almost guaranteed) that the father of her first child will not be the same as that of her later offspring. As Aristotle put it: 'This is the reason why mothers are more devoted to their children than fathers: it is that they suffer more in giving them birth and are more certain that they are their own'. The conflicts of interest involved, and the differences in the investment of each sex in their young, may help to explain certain strange patterns of inheritance.

Much to the surprise of geneticists the effects of a particular gene sometimes depends on whether it is passed on by mother or by father. This effect, 'genomic imprinting' as it is known, is quite different from sex linkage, as the genes involved may be on any chromosome. Each sex seems to stamp its personality on what it transmits. The DNA itself is not permanently altered, but its effects on those who inherit it depend on which parent it came from. A gene passed on by a father to his daughter differs in its impact from that of the same gene passed by her to her own children. The DNA is 'marked' as it is conveyed through sperm or egg and the mark reversed when the line of transmission changes from one sex to another.

Each embryo contains both maternal and paternal DNA. If we use the (rather dubious) metaphor that every gene acts in its own interests, it pays those that come from the father to extract as much as possible from the mother in which they find themselves, irrespective of any damage which this does to her or to subsequent children, as later offspring may well carry genes from a different father. The first father loses nothing by exploiting his mate as much as he can as she will bear no more of his children. The mother, in contrast, needs to ensure that further attempts to pass on her own heritage are not jeopardised by the avarice of her firstborn. The difference in behaviour of the same gene when transmitted through fathers or through mothers hence arises from paternal greed.

In mice, the genes that make the membranes through which the foetus feeds are more active if they come from the father; and the effect is so strong that the placenta itself has been described as a parasite forced on the mother by the father. Those passed on by the male parent also tend to increase the size of the tongue (which is, of course, used in suckling). Genes for human disease show the same effect. Some foetuses by accident inherit two copies of a gene that promotes growth. They become abnormally large only if both copies come from the father. In normal foetuses just the paternal copy is switched on as further evidence of the father's interest in his child extracting the most it can from its mother. Two rare genetic diseases (the Prader-Willi and Angelman syndromes) were once thought to be different as their symptoms are distinct. In fact they are due to the same mutation, a deletion of a short segment of chromosome fifteen which damages the gene involved in imprinting paternal identity . The differences depend on whether it is passed on by father or by mother. Children with Prader-Willi syndrome (whose abnormal chromosome comes from their father) suckle hard, become obsessively interested in food and are fat, while Angelman children (who receive the same structure from their mother) are thin or of normal weight, but have quite severe nervous symptoms, such as epilepsy, a tendency towards constant laughter, and a fascination with water. In the latter disease, the paternal copy of the gene is silenced in parts of the brain, perhaps explaining the mental illness.

Enthusiasts for conflict suggest that even a baby's cries are an attempt to manipulate its mother to provide more food and that the mother retaliates by secreting in her milk substances similar to those used as sedatives by doctors. Whether or not this is true, it is clear that once sex has evolved it has some unexpected effects on the lives of the creatures who practise it. Without sex there would be almost no evolution and no genetics. Our universal fascination with the subject may, one day, solve the most important sexual problem of all – why we bother in the first place.

Chapter Eight

CLOCKS, FOSSILS AND APES

The boundary between apes and humans was once far from clear. Lord Monboddo, a friend of Dr Johnson's, was convinced that 'The Orang Utan is as ardent for women as it is for its own females' and that the Malayans cut the tails off the offspring of such matings and took them as their own. 'From the particulars mentioned' – he wrote – 'it appears certain that they are of our species . . . though they have not come to the lengths of language.' Dr Johnson was not impressed: 'It is a pity to see Lord Monboddo publish such notions . . . in a fool doing it, we should only laugh; but when a wise man does it, we are sorry.'

A complementary foolishness is around today. Millions of Americans do not believe that humans are related to apes at all or even that the human species is more than a few thousand years old. Creationists are determined to stay ignorant. They deny that we evolved and are hence connected through our genes to the rest of life. President Reagan himself once said that 'Evolution is only a theory which is not believed in the scientific community to be as infallible as it once was . . . Recent discoveries have pointed up great flaws in it.' The creationist dogma bores, when it does not exasperate, biologists. As a result they have been less active in opposition than they should and the bigots have had some success, in the USA at least.

The written documents of history stop at the day before yesterday. The earliest texts come from the Sumerians. Records go back a little further in mythic form. Gilgamesh was King of the Sumerian city-state of Uruk in 2700 BC. The Epic which bears his name has some familiar features. It has a Garden of Eden, a Hero's descent to the Underworld (with a safe return) and a Flood. In the same way, the best of all evidence that humans did evolve and that they are members of the animal kingdom comes from our own ancient records. Without fossils, the portrait of our forebears can never be complete. A historian who knows only the modern world would find it impossible to infer the progress of, say, Turkey and the United States just from what exists today. All historians need documents from the past. To have confidence in their theories evolutionists must have the same thing.

Fossils authenticate the past. At one time, they were defined to be the work of the Devil, placed in the rock to mislead the faithful. Later came a last-ditch attempt to fit them into the Bible. Some fossil mammals seemed to be standing on tiptoe with their noses in the air when they met their end; proof that they had been overwhelmed by Noah's Flood.

Darwin was well aware of the power of the past. About one page in six of *The Origin of Species* deals with the fossil record of animals and plants. The fragments of their ancestors were central to his theory. For humans, Darwin had an enormous gap in his evidence. He knew little about the remains of our ancestors and there is scarcely a mention of them in his other great work, *The Descent of Man*, published in 1871. Although we now know a little more about the bones of our predecessors, the record of our own evolution is still very incomplete.

The first fossil to be recognised as a possible human ancestor was Neanderthal man, found in the Neander Valley in Germany in 1856. Such was the power of belief in those days that some dismissed the bones as those of an arthritic cripple or of a cossack who had died in the retreat from Moscow. A hundred years ago, a skull intermediate between humans and other primates was found. This was *Pithecanthropus erectus*, Java Man. The search for our birthplace was on and has continued ever since.

Palaeontologists still do not agree about where modern humans came from and where they went. The fossil record is so incomplete that a cynic might feel that the main lesson to be learned is that evolution usually takes place somewhere else. The origin of humanity has been claimed to be Asia, Africa and even the whole world at the same time. The human record has been studied as hard as any, but still has enormous holes. Even the best known deposits are sketchy. The area around Lake Turkana in East Africa is almost never off the television screen. Guesses about population size suggest that perhaps seventy million people lived there over its two and a half million year history. Remains of just a few hundred have been found, most as small fragments. The fossil record will never give us the complete history of our past, but it can give dates and places which genes can only hint at. It is worth glancing at the bones before staring at the molecules.

Just as with the genome, the biggest problem with the preserved record of the past is one of scale. Life began about four thousand million years ago. The journey from Land's End to John o'Groat's can again be used as a metaphor. Everywhere south of Birmingham is covered with primeval slime about which we know little. The first primitive land animals crawl ashore near Edinburgh. There are frogs in the Cairngorms and for thirty miles north of Inverness the landscape is infested with dinosaurs. The first primates appear near Wick, while our own species can look over the icy waters of the Pentland Firth from its birthplace a hundred yards from the northernmost shore of Britain. Recorded history begins on the beach, at high tide mark.

The journey through time needs milestones. Fossils can be dated in many ways. Some depend on the decay of radioactive materials as time passes. Others are more ingenious. Ostrich eggs were favoured as containers in the ancient world. The structure of their amino acids, like that of all tissues, is biased towards the left. Over the years, the amino acids decay into a mixture of left- and right-handed forms. To measure the ratio of left to right dates the shells and the people who used them. The oldest ostrich-shell containers are at Klasies River Mouth, a site in South Africa occupied by humans whose skulls are much like those of today. They are dated at a hundred and twenty thousand years old. The oldest outside Africa, found in the Israeli cave of Qafzeh, are twenty thousand years younger; and fifty thousand years ago the shells were used to make the first of all ornaments, some beads unearthed in Tanzania.

Our history stretches back to the same dawn as all other creatures; and, like them, we are improbable survivors from a past that has almost disappeared. There were several nasty moments on the road to *Homo sapiens*. For much of history, several species of pre-humans lived at once. Most went nowhere. As apes, we belong to the less successful branch of our family, for most of our kin disappeared fifteen million years ago as monkeys flourished. As primates humans claim allegiance to a group that thirty million years earlier left only a few twigs on a flourishing family tree; and as mammals we are the descendants of a rare and insignificant band of mouse-like creatures that cowered beneath the once thriving dinosaurs.

The history of mankind's earliest ancestors is obscure. Bones that look like those of primates – apes, monkeys and humans – appear around sixty million years ago. The first fragment of what may be an anthropoid (the group which evolved into monkeys, apes and humans) is about forty million years old. This creature was not much bigger than a rat; but within ten million years the group was well established from North Africa to China and Burma. The animals were small, not much larger than today's tarsiers and, like them, they probably ate insects and fruit rather than the leaves favoured by many of their modern descendants. A jawbone half that age from an early hominoid (the group which includes humans and apes) has been found in Kenya. By fifteen million years ago several species of ape roamed Africa and Asia. None was larger than a seven year-old boy and all had small brains and pointed faces.

One of the earliest direct precursor of modern humans appeared between three and four million years before today in the Laetoli beds of Kenya. *Australopithecus afarensis* is named after the Afar region of Ethiopia, the Biblical Ophir referred to in the story of Solomon and Sheba. The most famous specimen is 'Lucy', so named because the discoverers were playing the Beatles' 'Lucy in the Sky with Diamonds' at the time. She was less than four feet tall, with a small skull. *Australopithecus* bones show that, like today's chimpanzees and gorillas, they went in for knuckle walking. The earliest bones which look as if they belong to our immediate ancestors, the genus *Homo*, come from Kenya and are dated at about two and a half million years old. The first stone tools appear at about the same time.

It is as difficult to classify fossils as to define artistic styles. Because they evolve one into another it is pointless to draw a line to show exactly when, for example, impressionist painting changed into post-impressionism. A certain arbitrariness is bound to creep in. In palaeontology things are even worse, as few specimens are found and their discoverers have a natural tendency to grace each with its own name. Even what is needed for promotion to that august genus, *Homo* is in dispute. A pint or so of brain, a grasping hand and a few simple stone tools are the minimum entrance requirements; but the pass mark depends on the whim of the examiner (some of whom are choosy enough to exclude the oldest member, *Homo habilis*, and to downgrade him to a mere australopithecine).

However, most experts agree that there were at least four species of *Homo*: the first, *Homo habilis*, ('handyman') from around 2.3 million years ago to its disappearance some seven hundred thousand years later; the second, *Homo erectus*, emerging about 1.9 million years before the present, with the youngest reliably dated specimens at about 200,000 years old, and – in the end – our own species, *Homo sapiens*, which began to emerge about half a million years ago. *Homo habilis* is sometimes divided into two distinct species, *habilis* itself and *Homo rudolfensis*. *Homo erectus*, too, has been subdivided into two or more species; and the history of our species may be one of a series of close relatives who became extinct, leaving no trace of their presence. *Habilis* had a larger brain than its predecessors, its face jutted out less and for the first time there was a noticeable nose and a perceptible chin. An almost complete skeleton of an *erectus* boy has been found near Lake Turkana in Kenya. He had a brow ridge and a massive jaw, with long arms and legs. For much of the time more than one species of man-like beast existed at once. In Africa beasts which looked much like Lucy and her relatives lived for thousands of years alongside *Homo habilis*. For much of the time, several species of *Homo* may have lived together, a situation hard to conceive today.

Homo erectus was the first to escape from Africa and did so soon after it appeared. *Erectus* bones almost two million years old, mixed with those of sabre-toothed tigers and elephants, have been found in the the Republic of Georgia and this species soon spread, in modified form, to the Middle East, China, Java and Europe. Its remains include 'Java Man' and 'Peking Man' (whose bones disappeared during the Japanese invasion of China) which had a static existence with almost no change in the skull over its long history.

The first *Homo sapiens* – some of which look rather like *erectus* – emerged in Africa more than four hundred thousand years ago. They were robust and would appear distinctly hostile to modern eyes, although some had brains larger than the average today. A half-million-year-old skull that may belong to this group was found at Boxgrove in Sussex. Within a couple of hundred of thousand years such populations of 'archaic *Homo sapiens*' were found throughout Europe.

The Neanderthals, who flourished for a hundred thousand years before they sank beneath a wave of modern humans, had larger brains than our own (albeit on a heavier body), with large noses and teeth. Their remains have turned up all over Europe and the Middle East. and have been found as far east as Iraq, but not in Africa or elsewhere. Once, they were assumed to be on the direct line to ourselves; but fragments of DNA extracted from mitochondria show them to have been so distinct that Neanderthals mark yet another dead end on the road to humankind.

Around a hundred and thirty thousand years ago, the first humans of modern appearance (light build, thin skull, large brain and small jaw) appear in Africa. Their remains have been found from Omo-Kibish in Ethiopia to the southern tip of Africa, thousands of miles south. Quite soon after they emerged, modern humans began a relentless expansion. Many of the earliest sites are near the coast: but in those distant days the sea was lower than it is today and broad coastal plains stretched in front of those ancestral caves. Stone tools have been found on coral reefs in the Red Sea (which was dry a hundred thousand years ago) so that perhaps our early ancestors walked along a now-drowned shoreline from Africa to Indonesia, leaving their archaic relatives, who were still around at the time, inland.

These early modern humans had arrived in Israel, in the caves of Qafzeh and Skhul, by a hundred thousand years ago. Cro-Magnon man, the first modern European (who lived, like a sensible man, in the south of France) was there by forty thousand years before the present day. As in Africa, their antiquated relatives, the Neanderthals, held out for a time; and lasted in southern Spain for ten thousand years after the arrival of the newcomers.

This account of history is the 'out of Africa' model believed by most evolutionists. Some feel that humans emerged more or less at the same time over the whole world so that today's Chinese evolved from an ancient Chinese ancestor and Africans from a predecessor in their own land. The idea that the same species can evolve simultaneously in different places flies in the face of theories of the genetics of speciation. Some fossils might, perhaps, support the idea of local evolution. One, from the Han River in China, resembles *Homo erectus*, but has a flattened face which, to its discoverers, looks rather like that of a modern Chinese. Those in favour of local evolution make much of the 'shovel incisors' in fossil jaws from Asia. The teeth are scooped out at the back, as are those of some of today's Chinese. In some places in Europe, as well, a third of people have shovel incisors, so that this is not a forceful argument. So few fragments have been preserved that it seems that, too often, history is in the eye of the beholder. Africa was the centre in which most primates originated and there is no reason to suppose that humans came from anywhere else.

Another fossil controversy is the question of evolution by creeps or by jerks. Darwin felt that the origin of species was gradual and continuous. The past was no more than the present writ large. Because of the vast time available the enormous transformations which took place through the history of life could be explained by the slow and almost imperceptible changes that influences it today. His was a leisurely and Victorian view of the way the world worked; one of gradual and almost inevitable movement.

The opposing view (the theory of 'punctuated equilibrium' as it is known in its latest guise) has a more twentieth-century flavour. It sees evolution as boredom mitigated by panic. New species appear as the result of a sudden burst of revolutionary change. Between these historical disasters, life is tranquil. Punctuationists suggest that the origin of species has little to do with what happens to a species once it has originated and that the process of evolution today cannot tell us much about what went on in the past.

The greatest strength of this theory is its ability to annoy Darwinists. Hundreds of scientific papers have been written in support of or against the idea. One important problem is that of time-scale. What might appear an instant to a geologist can seem an eternity to a biologist. A 'jerk' between one species and its successor may encompass tens of thousands of years; nothing in terms of geological epochs, but more than enough generations to allow gradual evolution to make fundamental changes. Those opposed to creeping advance point out, quite fairly, that most species do not change at all through their evolutionary career, which is not what Darwin would have expected.

Whatever the merits of each doctrine, the human fossil record has so many gaps that there is just not enough information to tell whether humans evolved suddenly or slowly. The remains are so sparse that it is quite possible that relics of the lineage that led to the peoples of today are as yet undiscovered. Fossils are the best of all evidence that we did evolve, but cannot tell us much about how that evolution took place. They do show that the attributes which make us what we are arose bit by bit, seen first in a remote ancestor and reaching completion (if indeed they have) only in the past hundred thousand years or so. No single primate awoke one morning to find itself human.

The central problem in using the dead to recreate history is that we can never be sure that any fossil left a descendant. Our extinct predecessors are just that: extinct. This makes it difficult to work out their relationships to each other and to ourselves.

Another window has opened onto the past. Every modern gene descends from times long gone. The connections between humans and primates are preserved in the DNA of living animals. Darwin himself saw better ways to reconstruct history than to depend on the frozen accidents of fossilisation. His own claims about our predecessors used indirect evidence (such as a comparison of the anatomy of humans with that of apes). Nowadays this evidence is much more complete and has painted a new portrait of our ancestors.

Molecular biology is just anatomy writ small, plus an enormous research grant. To a geneticist, everyone is a living fossil and contains the heritage of his or her predecessors. Genes recreate the ancient world. 'Man' – Darwin said – 'still bears in his bodily frame the indelible stamp of his lowly origin.'; or, as W. S. Gilbert put it: 'Darwininian man, though well-behaved, is really just a monkey shaved.' Genetics allow us to search for who that shaved monkey may have been.

Bones show that humans are more related to apes than to monkeys and that their closest kin lie among chimps, gorillas and orang-utans. Anatomists once assumed that *Homo sapiens* must be quite distinct. Often, it was contrasted with these 'great apes'. We differ from them in obvious ways – brain size and hairiness, for example, and in other talents. Most people are righthanded and the patterns of breakage of stone tools suggest that our ancestors were the same. Individual chimps and gorillas may use one hand rather than the other, but about half the animals prefer left and half right. The human brain, too, is asymmetrical and it is more than a coincidence that speech and language are coded for on only one side.

It is hard to measure how much genetic divergence a difference in hairiness or handedness might represent and such comparisons are not much use as a test of the biological gap between apes and humans. DNA does a better job. Men, apes and monkeys share many genes. We vary both in the way we taste the world, and in how we see it. In some monkeys, many of the males are red-green colour-blind. Chimps have A and O blood groups, while gorillas are all blood group B. About a thousand distinct stained bands can be seen in the human chromosome set. Every one is also found in chimps. The main change is not in the amount of chromosomal material but in its order. Many of the bands have been reshuffled and two chromosomes are fused together in the line that led to humans. We have forty-six chromosomes in each cell, while chimps and gorillas have forty-eight. Pope John Paul in 1996 referred to an 'ontological discontinuity' between humans and apes. Perhaps that moment was marked by the fusion of ape chromosomes 1 and 2, with the human spirit coded somewhere on its amalgamated equivalent.

At the DNA level, too, (and forgetting issues of ontology) there has been little physical change. In one of the haemoglobin pseudogenes, the 'rusting hulk' of a gene (a structure that accumulates many mutations as it has no function), humans are about 1.7 per cent distant from both chimps and gorillas, 3.5 per cent from orang-utan and 7.9 per cent from rhesus monkeys.

To sort out man's place in nature we need to combine information from as many genes as possible. DNA hybridisation, crude as it is, does just that. The method depends on the extraordinary toughness of the DNA molecule and its overwhelming desire for togetherness; for each strand to pair with a sequence which matches its own.

When a double helix is heated up it separates into two individual strands, each of which bears a matching set of the four bases. As the liquid cools, the strands come together, A with T and G with C, to reconstitute the original paired structure. If DNA from two different species is treated in this way, some of the single strands from each one form a hybrid molecule that contains one strand from either species. The more related they are, the more similar their DNA and the tighter the fit. If the strands are very similar, they stay together even at high temperatures but, if they share fewer sequences, are less stable. The temperature at which the hybrid melts hence estimates how alike the two DNAs must be and gives a measure of their relatedness. DNA hybridisation has already sorted out some thorny problems of classification. Thus, it shows that the nearest relatives of New World vultures are storks, rather than the vultures of the Old World.

The results from primates are remarkable. Humans and chimps share 98.4 per cent of their DNA, rather more than either does with the gorilla. The orang-utan is not so related and New World monkeys even less so. Any idea that humans are on a lofty genetic pinnacle is quite wrong. For a series of functional genes, the DNA sharing between ourselves and chimps is as much as 99.5 per cent, with the gorilla further out on the same limb. A taxonomist from Mars armed with a DNA hybridisation machine would classify humans, gorillas and chimpanzees as members of the same closely-related biological family – indeed, he might count all three as so similar as to share the Latin name *Homo*, once seen as unique to ourselves.

Humans and chimps are not, needless to say, just minor variants on the same theme. Evolution involves more than overall change in DNA. The Hawaiian islands have more species of fruit fly than anywhere else in the world, with a vast diversity of form. One looks like a hammerhead shark, with huge protuberances on either side of its head. DNA hybridisation shows that this wild evolutionary euphoria is accompanied by almost no change in the genetic material.

Brains and behaviour are what separate humans from any other animal. Since the split from chimps, the brain has added about a thousand cells a year. The human brain is five times larger than would be expected for a typical primate of the same size. The genes involved lose their importance in a measure of average genetic difference; and the brain itself produces a whole set of intellectual and cultural attributes that appear once a crucial level of intelligence has been reached and are not coded for by genes at all.

Somewhere in that brain, or what it is thinking, is what makes us different. Although it shares most of its DNA with humans, no chimp can speak. Some claim that they can manipulate symbols in a primitive 'language' (and trained parrots can do almost as well). The attempt to show that apes might talk is one of the great blind alleys of behavioural research. Samuel Butler's comment on a Victorian attempt to teach a dog sign language makes the point: 'If I was his dog, and he taught me, the first thing I should tell him is that he is a damned fool!' To make too much of the shared DNA of chimps and humans is to be in danger of the same foolishness. Humans, uniquely, are what they think.

Whatever their limitations, shared genes say a lot about history. The biological differences between humans and their relatives come from the mutations that have taken place since the primates began to diverge. They can be used to guess at when the human family separated from the others: the more differences, the longer ago the split. If such genetic accidents happen at a regular rate, they can even be used as a 'molecular clock', which uses changes in genes to infer when two stocks last shared a common ancestor.

Molecular clocks depend on several assumptions, some of which may be justified. First, mutations must happen at a constant pace as the generations succeed. In addition, they should not damage their carriers (and, as most of those used are in parts of the DNA that do not contain any meaningful instructions, perhaps they do not). DNA errors accumulate over the years. Some are lost because, by chance, those who carry them do not reproduce, but are replaced as new ones come along. The genetic make-up of any line hence changes with time. The transformation of the inherited message is a clue as to when two species began to diverge. To date the split, there must be evidence from fossils (or from other sources such as the date of appearance of a barrier such as a mountain range) about when two extant members of the group last shared a common ancestor. To compare their genes sets the rate at which the clock ticks and makes it possible to work out the date of separation of others whose ancestors left no fossils.

Linguists use the same logic. As words are passed from parents to children errors creep in. Sometimes the changes are scarcely noticeable. In Shakespeare's *As You Like It* the court jester makes a speech which causes great amusement: looking at a clock he says 'Thus we may see how the world wags; 'tis but an hour ago since it was nine; and after one hour more 'twill be eleven; and so, from hour to hour, we ripe and ripe; and then, from hour to hour, we rot and rot; and thereby hangs a tale.' Just why this should be so funny is lost on modern audiences; unless they realise that in Shakespeare's time the word 'hour' sounded almost the same as the word 'whore'. Such errors can be used to date manuscripts. They were copied by hand, often by people who had little idea what they meant. As copy followed copy, more and more mistakes crept in. The number of inaccuracies gives a good idea as to when any version of an original was in fact written.

Such changes are small, but can make a big difference. Languages as distinct as Bengali and English are related. They owe their existence to the accumulation of tiny changes to a common ancestor spoken long ago. Take the word for kings and queens. In Sanskrit, this was *raj*, in Latin, *rex*, in Old Irish *ri*, in French *roi*, in Spanish *rey* and in English, royal. Different transmission errors took place in the pathway to each language. If we know the date of the split (as we do from the literary fossils known as books) we can make a linguistic clock. In Europe, this ticks at a rate which means that two languages share about eighty per cent of their words a thousand years after they divide. The language timer is an imperfect one: some words hardly change while others shift more quickly. Nevertheless, it can be used to trace the origin of modern tongues although their ancestral speakers are long dead.

The idea of a molecular clock driven by mutation is elegant and simple. As usual, the more we learn the worse it gets. It speeds up and slows down, and ticks at different rates for different genes. A similar confusion led the nineteenth-century Linguistic Society of Paris to ban discussion of the origin of languages. There have been some spectacular failures by molecular clockmakers to see the biological wood for the evolutionary trees. Even so, they have had some triumphs and these can be used to date human evolution. Several cautions are called for. The primate fossil record is so patchy that it is difficult to find firm dates with which to set the clock. Fossils and genes suggest that the primates began to diverge about sixty million years ago, the line to baboons had split off by twenty-five to thirty million years ago and that to orang-utans by twelve to sixteen million years before the present. They say nothing about the date of the split between humans, chimps and gorillas. A molecular clock based on the genes of our relatives suggests that this division took place five to seven million years before the present, with the divergence of the gorilla line before that of those to chimps and to humans.

The last common ancestor of chimps and humans hence lived about three hundred thousand human generations ago. *Homo sapiens* is a recent arrival even in the history of the primates, let alone in the four thousand million-year pedigree of life itself. The molecular clock suggests that the Philippines tarsier (an unremarkable small brown monkey) split from the western tarsier (which is almost identical in appearance) at about the same time as our divergence from chimps, showing just how quickly our ancestors must have evolved.

When – or how – the attributes which separate humans so absolutely from any other creature first appeared is not a question for biology. Perhaps the best we can do is to agree with Keats that we are all 'twixt ape and Plato' and leave it to individual preference just where on that long road we place ourselves.

Chapter Nine

TIME AND CHANCE

The Good Book points out, in Ecclesiastes, that 'The race is not to the swift, nor the battle to the strong . . . but time and chance happeneth to them all.' Evolution is all about change with time, but how things evolve is often a matter of chance. The nature of inheritance means that random events are bound to direct our genes as the generations succeed one another. As a result, much of the human condition is shaped by accident.

The importance of chance in evolution was noticed by the English cleric Thomas Malthus. He became interested in the history of the burghers of Berne and followed the records of their names over several centuries. To his surprise, many of the surnames present at the beginning of the period had gone by the end, although the number of burghers stayed about the same. Francis Galton showed why.

A surname is rather like a gene; it passes from father to son. Every generation there is a chance that a particular father does not have a male child. Perhaps he has just daughters (who lose their names on marriage) or no children at all. His name is then lost from his family line. If he has no children the name will go at once. It is also more likely to become extinct if he has a small family, as a limited brood will quite often consist of daughters alone. If, in a closed community like that of the Bernese bourgeoisie, this process goes on for long enough, more and more names will disappear as the years pass. Given enough time, just one surname will survive. Everyone will carry the same inherited message (at least as far as their name is concerned). In addition, the community will be more inbred than before, as the only mates available will be people who share the same name, all of whom descend from a common ancestor.

Just the same happens to genes. Perhaps, among the mediaeval Bernese bourgeoisie, there was a rare gene – an unusual blood group, for example. Because Berne was a small town, few people carried it. If none passed it on (because they had no children, or because by chance the gene did not get into the sperm or egg that made the next generation) then it was lost. On the other hand, the carriers might, again by chance, have had more children than the rest, in which case the variant became more common. In either case, its frequency altered (which means that the population evolved) in a manner that depends only on the accidents of time.

Einstein once said that 'God does not play dice.' He was wrong: for genes, God does. What number comes up has nothing to do with the DNA involved. That raises an almost theological issue. Is it their own fault that genes, and those who carry them, are damned – or do they perish at random because of simple bad luck?

Such evolution by accident is known as genetic drift. The process has been important in our own past. *Homo sapiens* was until not long ago a rare species that lived in small bands. Until a few tens of thousands of years before the present there were no more people worldwide than live in London today. The few tribal peoples to have survived hint at what society was like.

Until the 1970s, when their lives were destroyed by gold-miners and loggers, about ten thousand Yanomamo Indians lived in a hundred scattered villages in the rain-forests of southern Venezuela and northern Brazil. They called themselves 'the fierce people', with good reason.

About a third of all male deaths were due to violence, often in battles between the villages and, the Yanomamo believed, many more to malevolent magic.

Their society was not robust enough to allow groups of more than eighty to a hundred people, including around a dozen young adult males, to stay together. Any larger band tended to split. The splinter group moved away to found a village somewhere else. As a result, the Yanomamo existed for their whole history (which stretches back in some form to the earliest Americans twelve thousand and more years ago) as a series of small communities in constant conflict.

Most hunter-gatherers may have lived like this. The ancient Siberians who hunted mammoths made houses from their bones. The size of their bony villages suggests that each group, like today's Yanomamo, consisted of a few score people. One odd fact about modern society may also be a hint about the size of ancient groups. Most team efforts involve about the same number of individuals. There are nine members of the US Supreme Court, eleven on a football team, twelve on a jury – and Jesus had twelve disciples. Each Yanomamo band has, curiously enough, about a dozen healthy adult males. Is the difficulty of reaching consensus in a larger group a relic of earlier times? Most people can identify about twelve others whose death would cause them anguish. Aristotle himself pointed out that it is impossible to love more than a few. Could this be a clue (albeit a feeble one) about the size of ancient communities?

Just as for surnames, random genetic change takes place more easily in small populations, when few people bear a particular gene. Then, all or most of the carriers may, by chance, fail to transmit it. In a larger group, a variant may be rare but will be borne by enough people to ensure that at least one will pass it on.

Strange things befall genes in small populations. Again, surnames show what can happen. Their evolution is easy to study, as it needs no more than a telephone book, and names are preserved for centuries in marriage records. The world has about a million surnames. Those in China are the oldest and date back to the Han dynasty two thousand years ago. In contrast, Japanese surnames go back only a century or so, when they were ascribed by order of the authorities. Various complications face those who study them. For example, in many places the same name (like my own, Jones, which means 'son of John') appeared independently many times. In some societies, such as those of Spain and Russia, the system breaks down as children take the name of their father and the 'surname' changes each generation. The same was once true in Wales. A boy would take the name of his father and more distant ancestors, each prefixed by the term 'ap', or 'son of'; and – the more the names, the more respected the family. Remnants of the system exist in modern Welsh surnames such as Pugh (son of Hugh), Price (son of Rhys) and Parry (son of Harry). In most places this practice has almost gone.

The telephone book in a long-settled part of the world (such as the mountainous country around Berne) shows that villages just a few miles apart each have a distinct set of names, with, in some villages, almost everyone a bearer of the same one. Within each isolated hamlet there has been an accidental loss of names as, by chance and over the years, some men have had no sons. Because the effect is random, different surnames have taken over in each place. The process may be helped by each village having been founded by a group which had, again by chance, its own characteristic set of last names. It is not the case that within a village one family label is somehow better than the others. Instead, its prevalence reflects the errors of history.

The genes of isolated populations are much the same.

Adjacent Yanomamo (and even Alpine) villages have rather different frequencies of blood groups and other variants. In Alpine villages, blood group frequencies diverge to just the extent predicted from what marriage records say about their size since they were founded. They have evolved by accident.

In large modern cities such as Berne the picture is quite different. The phone book contains thousands of names, none of which is particularly common. Again, the rules of chance and time are at work. Cities contain so many people that it is unlikely that any name, or any gene, will go extinct just because its few carriers fail to pass it on. Such places attract immigrants, so that new names (with their associated genes) come in all the time and the population becomes more diverse. A simple but effective way to measure how distinct a community might be is to count the number of surnames in relation to the number of people. If more or less everyone has a different label then the community is open to migration from many places and is, in effect, so big that accident is unimportant. A glance at the New York telephone directory compared to that of, say, Oslo shows at once that the two have had different histories. The USA has a higher proportion of all global names than anywhere else. That reflects its chronicle of immigration from all over the world.

Shared names mean shared ancestors which in turn means shared DNA. A population in which many people carry the same gene (or the same surname) because they have inherited it from a common ancestor is said to be inbred. To some extent we are all inbred as we are all to some degree related. Everyone has two parents, four grandparents and so on. If all were unrelated, the number of ancestors would double each generation to give an absurd number of ancestors within a few centuries. In fact, related people married, and the lines of descent have merged and blended. As a result we all have many ancestors in common.

Perhaps the most inbred individual ever recorded was an aristocrat, Cleopatra-Berenike III, aunt of the Cleopatra enamoured of Anthony. She may have had identical copies of half her genes because they descended from a single ancestor. As the ancient Egyptians saw the pharaohs as their gods' posterity they were anxious to keep the deities' blood-line as pure as possible with mating among relatives (sometimes, even, between brother and sister). The story is confused by difficulties in reading the hieroglyphs showing degrees of pharaonic relatedness.

Levels of inbreeding vary greatly from place to place. The incidence of marriages between people with the same name is quite a good indicator. This was noticed by George Darwin, son of the more famous Charles (who married his own cousin). He estimated from surnames that the proportion of cousin marriages (the closest legal form of inbreeding) among British aristocrats, by definition a small and exclusive group, was about four and a half per cent – more than twice that in the general population of his time. The pattern of family names shows that the British population as a whole is, on the average, more outbred than much of the rest of Europe. Even in remote and rural East Anglia, just one in fifty of those present at the end of the eighteenth century had been there in the seventeenth, evidence how much movement there had been in comparison to Switzerland or Italy.

A small village offers little choice when it comes to picking a spouse. As a result, relatives marry and the population becomes inbred. Sometimes the married couple have each received a copy of a harmful recessive gene from their common ancestor. As a result their children are at increased risk of having two copies. George Darwin found that Oxford and Cambridge oarsmen, a healthy group, were less likely to have issued from a cousin marriage than were their more indolent peers.

There are constraints on how close a relative one may marry. Brother with sister is forbidden everywhere but even first cousin marriages may be illegal (as in most US states in the nineteenth century and in Cyprus today). This social imperative may have arisen, in part at least, from a fear that the children might be less healthy. As childhood mortality was in any case so high when the taboos were formulated (so that a small increase because of genetic disease would not be noticed) perhaps they have no biological basis at all.

The death rate does increase and development slows in the children of close relatives. Cousins share a grandparent in common. If he or she carried a harmful recessive (as almost everyone does) their children and grandchildren are more likely than average to inherit two copies. In some Japanese villages before the Second World War, up to a third of all marriages were between cousins. The huge survey of the population of Hiroshima after the atom bombs showed that the children of cousins walked and talked later than others and did worse in school. Part of this was due to the relative poverty of their parents but part reflects their heritage. The same is true in India and in Pakistan, where up to half of all marriages are still between cousins or between uncle and niece. The picture is confused here because such marriages tend to retain wealth within the family and to increase the number of children the parents can afford. Nevertheless, these too survive less well than the children of unrelated parents. First-generation Pakistani immigrants to Britain are also rather inbred. Just one birth in fifty is to such parents, but about five per cent of all inborn disease among British children is to those with Pakistani parents.

It is important not to overstate the dangers of inbreeding. Parents who are cousins have rather more than a ninety per cent chance of a perfectly normal baby, compared to more than ninety-five per cent for unrelated parents. Inbreeding has an effect, but it is dwarfed by the improvements in child health which have taken place in the past few decades. The map of human genetic diversity, based as it is on thousands of points across the genome also gives an insight into inbreeding: a child of a marriage between a couple with a common ancestor is likely to have double copies of long sections of identical sequence. In populations known from the records to share ancestors in common, this is often the case; but, quite often, children not otherwise known to be inbred also show the same decrease in variation in parts of their genome. The effects of forgotten inbreeding long ago can, it seems, persist for many generations.

In part because of this effect, isolated populations often show high frequencies of inherited abnormalities which are rare elsewhere. Most of the gypsies of South Wales belong to one extended kindred and half their marriages are between relatives (which makes them one of the most inbred peoples on earth). One Welsh gypsy in four carries a copy of the gene for phenylketonuria, which is four hundred times more frequent in this group than in Wales as a whole. A long history of social and sexual isolation has had an effect on their genetic health. Other close-knit family groups, such as the Bedouin of Israel, Jordan and neighbouring countries, may show high levels of inbreeding with, in some places, more than half of all marriages among cousins (and a concomitant local increase in diseases such as inborn deafness). Attempts by geneticists to promote outbreeding on health grounds have had limited success.

The effects of marriages of relatives may be subtle. A few women suffer from recurrent abortion. They often become pregnant, but the foetus is lost. The problem is found among the Hutterites, a religious group who originated in the Tyrol in the sixteenth century. In the 1770s, they moved to Russia, where they flourished and multiplied ten-fold from their original community of a hundred or so. A century later, bigotry was renewed, and the Hutterites migrated to America. All thirty thousand alive today, many of whom live in South Dakota, trace their descent from fewer than ninety founders and nearly all marry within the group. Over the years they have all become quite close relatives, and the more inbred a Hutterite woman might be the longer the interval between her children. Hutterite women who find it difficult to have children share, it transpires, a high fraction of their genes with their husbands. This may reflect the malign effects of inbreeding on the embryo.

In lower animals, genetic variation on the surface of cells determines whether a sperm is allowed to fertilise a particular egg. If the two cells are too similar, then fertilisation fails. Perhaps this is why the complicated system of genetic identification on the cell surface evolved in the first place. The repeated failure of pregnancy in genetically similar husbands and wives may be a remnant of a method of ending fertilisations which arise from the attentions of too close a relative. Spontaneous abortion, perhaps in the first few weeks of pregnancy, kills them off.

Mice have the mechanism in more dramatic form. Females can tell from scent how close a relative a male might be. Given the chance, they avoid mating with their brothers. What is more, if a mouse pregnant by a relative is offered an unrelated male (or even the scent of his urine) she aborts and mates with the new partner. The genes responsible for mouse scent are linked to those that control cell-surface variation.

Among the Hutterites, too, married couples are less simi-

The Language of the Genes

lar to one another for certain genes in the immune system than are pairs who are just friends. The genes involved are related to those which drive sexual choices in mice. Perhaps, quite unconsciously, most Hutterites – and most people – fall for someone with a set of identity cues different from their own. What is more, they are keenest to avoid a partner whose genes are too much like their mother's: the Hutterite mother is to be avoided as a role model in the choice of a wife. Just how the mechanism works, no one knows, but scent may be involved somewhere.

Accidental genetic change is close to how God might play dice. Statistics is needed to study it. Population genetics is infested with mathematics, much of which is incomprehensible even to population geneticists. It is, nevertheless, unavoidable. The importance of random change depends on the size of the population. It is not enough just to know the number of people around today. What is important is its average size since it began; after all, a large town may once have had just a few ancestral inhabitants. What is more, a special kind of average is needed. This pays particular attention to episodes of reduced numbers. Like so many ideas in evolution, the idea of the 'harmonic mean' comes from economics. Think of a village in ancient times, with one rich squire and many hungry peasants. Perhaps the fifty poor peasants each had an average income of a hundred pounds a year, while the squire gloried in a million. The average income was nineteen thousand pounds, which is a rather pointless statistic for anyone interested in rural reality. The harmonic mean income, in contrast, was a hundred and two pounds, which is a better reflection of what society was actually like.

The same logic applies to populations which change in number. Thus, the average size of a population whose size in succeeding generations is 1000, 1000, 10, 1000, and 1000 is 802 but its harmonic mean size is only 48. Any population bottleneck – ten individuals, in this case – has a dramatic effect that can persist for many generations.

To measure the real size of a population involves other subtleties. Variation in the number of children produced by each person means that its effective size may be less than first appears. Many tribal populations (and perhaps most ancient societies) show big differences in reproductive success, most of all among males. A few Casanovas monopolise the females, while lots of reluctant celibates do not get their fair share. Freud, in *Totem and Taboo* (delightfully subtitled *Some Points of Agreement between the Mental Lives of Savages and Neurotics*) built his theory of psychoanalysis on this: a supposed time of a primal horde led by a dominant father with sexual rights to all the women. His sons killed and ate him, inheriting the Oedipus complex which has been such a nuisance ever since.

Many societies do have a rather Freudian structure. In one Yanomamo village, four of the old men had 41, 42, 46 and 62 grandchildren respectively, while twenty-eight had only one grandchild and many more had none. Women, on the other hand, each had about the same number of descendants. A simple count of the men would much overestimate the real population size. From evolution's point of view, many of them might just as well not be there.

All populations have a history. The iron rules of chance mean that any episode of reduced size – a population bottleneck – will have a prolonged effect. From earliest antiquity humans have been colonisers, first as they filled the world from their African home and later as economic pressure drove people to conquer new lands. The emigrants were a small group, a tiny sample of the people left behind. The new colony may grow into millions, but all its inhabitants carry only the genes of the founders. As there were so few pioneers, the new population may be, by accident, quite different from those who stayed at home.

This 'founder effect', as it is known, is important throughout evolution. Darwin's first port of call on the *Beagle* voyage was the island of Madeira. He commented on how different its snails were from their European ancestors. This difference became even more conspicuous when he began to look at the birds and tortoises of the Galapagos. Perhaps, Darwin thought, the accidents of history, with chance colonisations of each island, helped to explain why archipelagos were natural laboratories for evolution.

The quirks of colonisation have been just as important in our own past. Ironically enough, the best example of evolutionary accident comes from not an escape but a return: the Afrikaners' journey back to their ancestral continent after an absence of more than a hundred thousand years. They began their migration from Europe in the 1650s. The pioneers brought with them a lasting legacy. It included more than Calvinism and bigotry. The surnames and the genes of their descendants are still a bequest from the first migrants. The three million Afrikaners in South Africa all derive from a small group of settlers, some of whom were so enthusiastic in their fecundity as to leave tens of thousands of descendants today. A million Afrikaners share just twenty names (Botha being one). This fits what history tells us about the number of immigrant families. Even today, half the surnames arrived before 1691 and the other half before 1717.

The migrants also brought, quite unawares, some rare genes drawn by chance from the people of Holland. One of the partners in the marriage of Gerrit Jansz and his wife Ariaantje Jacobs (who was one of a group of girls sent from a Rotterdam orphanage in the 1660s) must have carried a copy of the gene for a form of porphyria. This disease (which is related to that which may have afflicted George III) is due to a failure in the synthesis of the red pigment of the blood. Sometimes, light-sensitive chemicals are laid down in the skin. Here, they react with sunlight and produce painful sores. In certain forms of porphyria hair grows on exposed areas. Sometimes the waste material accumulates in the brain and leads to mental disorder. Part is excreted in the urine, to give a characteristic port-wine, almost blood-red, colour. Werewolves – creatures that come out at night, howl and drink blood – may have begun with the porphyria gene.

The South African form is mild but became important when barbiturate drugs were used in the 1950s. Carriers of the gene suffered pain and delirium when they took them. Porphyria is rare in Europe, but thirty thousand Afrikaners bear it. Johannesburg has more carriers than does the whole of Holland. All descend from one member of the small population of founders that grew in numbers to produce today's Afrikaners. Because it is so common in one family, porphyria in South Africa is sometimes called 'van Roojen disease'. A gene and a surname tell the same story.

The founder effect can be seen again and again among the descendants of those who colonised the world from Europe. Sometimes, the settlements are isolated by miles of ocean. Tristan da Cunha, a tiny island in the South Atlantic, was settled by a garrison sent to guard Napoleon, then in exile on St Helena. A few soldiers stayed on after the guard was withdrawn. They obtained wives by advertising, and a few shipwrecked sailors and others joined the community over the years. It went through a second bottleneck when several men drowned in a fishing accident and some families moved away, with the advice of a gloomy pastor. Now, the island is still a week's journey by ship from the mainland, but a few hundred people can stand the isolation. Again, they share names, seven altogether, and those – Bentley, Glass and Swain – of three of the first founders, are still common. Just five lineages of mitochondrial genes exist, and the island has its own genetic abnormality, a hereditary blindness brought by one of the original wives.

Some migrant communities are isolated by social rather than physical barriers. The United States has many religious groups whose founders emigrated to avoid persecution. They have grown into large populations which exclude outsiders. The Pennsylvania Amish have a unique inheritance. Almost a hundred babies have been born with six fingers and restricted growth, a condition almost unknown elsewhere. Every one of the affected children descends from Samuel King, a founder of the community.

To trace the movement of a gene around the world also shows the importance of chance. Huntington's Disease is relatively common among Afrikaners. Most cases descend from a Dutch man or his wife who emigrated in the 1650s. All copies on Mauritius are the legacy of a French nobleman's grandson, Pierre Dagnet d'Assigné de Bourbon, and more than four hundred patients in Australia have inherited their gene from a British immigrant, Mrs Cundick. Wales has a patch of the disease in the Sirhowy Valley, around the house of a mason who settled there in the nineteenth century and who must have carried the Huntington's gene. The largest kindred in the world (which was used to map the gene) is in Venezuela around an arm of the sea called Lake Maracaibo. Ten thousand descendants of one Maria Concepcion, who died in about 1800, have been traced. Four thousand either have the illness or are at a high risk.

Such accidents of colonisation must have happened again and again as humans spread across the world. Even without a written history, the surnames of the Afrikaners make it possible to estimate how many people were in at the beginning, three hundred and more years ago. Genes can do the same job. Patterns of variation show how many people founded a population, or whether it went through a bottleneck in the distant past.

Inherited diversity shows clear global patterns. Africans are more variable than are the rest of the world's peoples. Their cell-surface antigens (the cues recognised by the immune system) show about twice as much variation as do the equivalent genes in Europe, and many of its variants are unique to Africa. Africans are more variable for blood groups, proteins and DNA sequences as well. For mitochondrial DNA, the average difference between two Africans is twice that found elsewhere. Venezuelan Indians, in contrast, whose ancestors were near the end of the long history of movement across the world from Africa, have almost no variation in their mitochondrial DNA.

The decrease in diversity outside Africa, humankind's native continent, may be because genes were lost as small bands of people moved, split and founded new colonies in the trek across the globe. Just as for Afrikaner surnames the number of variants dropped each time a new colony was founded. The high levels of diversity among Africans is evidence that *Homo sapiens* has been in that continent for longer than anywhere else. Its decrease at the tips of the evolutionary branches in South America and Polynesia shows how human evolution was driven by chance as the migrants passed through a succession of bottlenecks.

A comparison of the genes of Africans with those of their descendants elsewhere in the world makes it possible to guess at the numbers involved in those early colonisations. The order of bases along a short length of DNA is in some ways a 'genetic surname', a set of inherited letters which pass together as a group down the generations. The name written in nucleic acids around one of the haemoglobin genes has been looked at in detail worldwide. The results are quite unexpected.

All populations outside Africa, from Britain to Tahiti,

share a few common sequences. Africa itself has a different pattern of distribution. Just like the names in the Johannesburg telephone book compared to that of Amsterdam, the shift in pattern from the ancestral continent to its descendants may be a relic of a population bottleneck at the time of migration – this time from, rather than to, Africa. We can do some statistics (and make quite a lot of guesses) to work out the size of this hundred-thousand-year-old group of emigrants. They show that the whole of the world's population outside Africa may descend from fewer than a hundred people. If this is true, non-Africans were once an endangered species.

Science has two cultures: one (to which most scientists belong) uses mathematics and the other understands it. Such guesses about ancient population bottlenecks demand statistical acrobatics. They also depend on one crucial, and perhaps quite mistaken, assumption; that the genes involved do not alter the chances of survival or of sex. Molecular biologists tend to assume that small changes in the structure of DNA are unimportant. It is just as possible that they do have an effect on fitness. If, for example, Africans have more variation on the surfaces of their cells because it helps to combat disease, then to claim that a reduction elsewhere is due to an ancient bottleneck is simply wrong.

Any attempt to reconstruct the distant past is bound to suffer from ambiguities such as these. Genetics has not yet revealed just how many Adams and Eves there may have been, but shows that much of the human condition has been shaped by accident: an observation that might at least instil a certain humility into those whose genes have defeated the laws of chance by surviving to the present day.

Chapter Ten

THE ECONOMICS OF EDEN

Renaissance painters on religious themes had a problem: when they showed Adam and Eve, should they have navels? If they did, then surely it was blasphemous as it implied that they must have had a mother. If they did not, then it looked silly. Although some compromised with a strategic piece of shrubbery, that did not resolve matters. And where was the Garden of Eden? Various theories had it in Israel, Africa and even the United States. When it existed seemed obvious because to add up the ages of the descendants of the primal couple as given in the Bible set the start of history as 4 October 4004 BC.

The reason for leaving Eden was also clear. Its inhabitants had, with the help of an apple, learned forbidden truths, and as a punishment were forced out into the world. No longer could they depend on a god-given supply of food falling into their hands. Instead, they had to make a living. The first economy was born.

The escape from Eden – the colonisation of the Earth – showed how genetic change is linked to economic development. Economics is often seen as a kind of enlightened self-interest. The desire to increase one's own wealth may, as Adam Smith has it, be the invisible hand which is at the foundation of all social progress. The same argument is used by some evolutionists. Genes are seen as anxious to promote their own interests, even at the expense of their carriers. In its most naive form, this view of life is used to explain (or at least to excuse) spite, sexism, nationalism, racism and the economic and political systems that grow from them.

Theories of economics and of evolution have obvious ties. Darwin was much influenced by the works of Malthus, who had been disturbed by the new slums of the English cities of the eighteenth century. In his *Essay on the Principles of Population* Malthus argued that populations will always outgrow resources. That notion led Darwin to the idea of natural selection.

Karl Marx, himself a denizen of one of the most congested of London districts, was just as impressed by the dismal conditions of the new proletariat. He sent Darwin a copy of *Das Kapital* (which was found unread after his death). Marx, in a letter to Engels three years after *The Origin of Species*, went so far as to say that 'It is remarkable how Darwin recognises among beasts and plants his English society, with its division of labour, competition, opening up of new markets, inventions, and the Malthusian struggle for existence.' Engels took it further. In his essay *The Part Played by Labour in the Transition from Ape to Man* he argued that an economic change, the use of hands to make things, was crucial to the origin of humans. If one substitutes the term 'tools' for 'labour' his views sound rather like those of modern students of fossils.

Genetics shows that much of evolution is, as Engels said, linked to social advance. However, far from society being impelled by its genes, social and economic changes have produced many of the genetic patterns in the world today. Every technical development, from stone tools on, has led to an evolutionary shift and to biological consequences that persist for thousands of years. Society – and most of all the economic pressures that cause people to move – drives genes, rather than genes driving society. Relentless expansion is at the centre of human evolution: in Pascal's pessimistic words, 'All human troubles arise from an unwillingness to stay where we were born.'

Fossils show that almost as soon as they evolved, humans began to migrate. Why our ancestors were so restless, nobody knows. Technological progress may have been involved, as the emergence of modern humans coincided with improvements in stone axes and the like (although tools had been made for at least two million years before the great diaspora).

Perhaps climatic change was as important. The Sahara Desert was once a grassy plain and Lake Chad a sea bigger than the present Caspian. Both dried up about a hundred thousand years ago, so that food shortage may have driven man out of Africa. A microcosm of that process is taking place at the southern edge of the Sahara. As the rains fail, the desert has spread into the Sahel and migrants are on the move.

The earliest economies had a simple foundation. People used what nature provided, until it ran out. The world is filled with fossils of large and tasty animals that were driven to extinction soon after humans arrived. In Siberia, so many mammoths were killed that the hunters made villages from their bones. In Australia, too, there was a shift from forests to grasslands as the immigrants burned their way across the continent. The record of destruction is preserved in the Greenland ice-sheet. The snows which fell tens of thousands of years ago retain the soot and ash from gigantic forest fires set by our ancestors.

New Zealand was not colonised until the time of William the Conqueror. For a few years there flourished a culture based on the exploitation of a dozen species of moas, giant flightless birds. The ritual slaughtering grounds where the birds were killed (and where half a million skeletons have been found) are still around. The birds themselves are not. In Europe, too, whole faunas went not long ago. Humans did not reach Crete, Cyprus and Corsica until around ten thousand years before the present. Before then they had some extraordinary inhabitants; pygmy hippos, deer and elephants, and giant dormice, owls and tortoises. Soon after the arrival of the first tourists, all were gone, and the burnt bones of barbecued hippos are scattered among the remnants of the earliest Cypriots.

The common large mammal in Europe and the Near East at the time when modern humans moved from Africa was one of their own relatives, Neanderthal Man. He had lived there quite happily for two hundred thousand years. Many Neanderthals found homes in the dense forests of southern France. Some had an economy based on hunting reindeer, with settlements concentrated around their migration routes. The cave of Combe Grenal in Perigord contains tens of thousands of Neanderthal stone tools. Their culture was, in its own way, sophisticated; but it did not progress and showed no real change for a hundred thousand years. Tools in Britain and the Middle East look almost the same. Those who made had little interest in exploration and never made boats, so that the delights of the Mediterranean islands (hippo-infested though they were) remained unknown. Neanderthals were the first conservatives.

Soon after the invasion of Europe by our own direct ancestors, they disappeared. Why, we can but guess. The guesses range from genocide to interbreeding. The first is unlikely. In France, at the cave of St Cesaire, Neanderthals and moderns lived close to each other for thousands of years. The second is probably wrong. If there had been sex between the indigenous population and the invaders, then modern Europeans would be expected to retain genes from this distinct branch of the human lineage and to have genes distinct from those of today's Chinese or Indians, whose ancestors never met a Neanderthal, let alone mated with one. They do not.

Perhaps economic pressure did away with those ancient conformists. For most of history, Africa was the most advanced continent. Africans made sharp blades while Europeans had to manage with blunt axes. There was a period when Neanderthals seemed to pick up some of the new technology, but it did not last. The first modern Europeans were found in 1868 during railway work, in the Cro-Magnon shelter at the Perigord village of Les Eyzies. Cro-Magnons looked much like modern Europeans. They (and their immediate predecessors the Aurignacians) had a sophisticated hunter-gatherer economy and made a variety of tools. Their cave art reached its peak around forty thousand years ago. The moderns had tools made of bone and ivory when their relatives still were satisfied with stone. They were better at exploiting what was available, so that their populations grew faster. That drove Neanderthals (and their genes) out. The last known skeletons are from St Cesaire. They died more than thirty thousand years ago.

Simple as it was, the Neanderthal economy held our ancestors at bay for a long time. The moderns reached Australia before they filled Europe. Competition from its indigenous inhabitants may have kept them out.

Most of the globe was populated at some speed after humans left their natal continent. The first Australians arrived about sixty thousand years before the present. The earliest remains are in two sites in Arnhem Land, in north Australia, which contain stone tools and ochre paints in a sandy deposit. The sites are close to the shore and perhaps to the point where humans arrived from the north. Soon, its inhabitants had complex tools and fishing nets and were economically as well developed as the rest of the world.

For much of its history Australia was joined to what is now New Guinea by a land bridge. It disappeared just seven thousand years ago. Tasmania was also part of

Greater Australia. That great continent, Sahul, has always been separated from Asia by a deep trench. The first Australians must have crossed at least ninety kilometres of water to reach their new home.

The passage may have been difficult, but the genetics of today's aboriginals suggest that it was made by many people. Native Australian DNA, like that of Papuans, is quite diverse. There must have been many founders, with several incursions into the continent. Once they got there, the new inhabitants found their home congenial and, at least in the tropical north, tended to stay put. As a result, in today's Papua New Guinea, local populations are quite different from each other, with distinct 'clans' of mitochondrial lineages, each limited to a few remote mountain valleys. Their denizens stayed isolated until the first Europeans reached the interior half a century ago. They were in their own way advanced, and cut down trees to allow the tastier plants beneath them to grow. Hidden in their fastnesses for tens of thousands of years they remained insulated from the economic strife and the waves of movement that affected the rest of the world.

At the other end of Sahul, rising sea-levels soon marooned the inhabitants of Tasmania. They remained in ignorance of the world outside until the arrival of Europeans, in the eighteenth century. Nothing is known of the Tasmanians' genes, for a simple reason. They were driven to extinction (and sometimes hunted down) by ambassadors of the modern world's economy. There was a sordid episode in anthropology when the Tasmanians were regarded – absurdly – as the elusive 'missing link' between humans and apes and the museums of the world quarrelled over the bones of the last survivors.

Human traces show that even remote Pacific islands (such as Manus Island in the Admiralty group, three hundred and fifty kilometres from the nearest land mass) were occupied twenty-eight thousand years ago, so that by then it was possible to make substantial voyages. The genes of present day Melanesians, those from the islands north and east of Australia, still resemble those of the ancient populations in the Papuan highlands. They are the descendants of these ancient voyagers.

The Polynesians who occupy the rest of the Pacific are quite different and got there much more recently. Hawaii and Easter Island were reached only a couple of centuries after the birth of Christ. In the far Pacific, islands separated by thousands of miles of ocean are not at all distinct in their genes, proof that water is a less effective barrier to movement than is land.

Almost all the peoples of the distant Pacific carry a small change in their mitochondrial DNA. Nine letters of the message are missing. This deletion has spread through the whole of Polynesia from Fiji to New Zealand. In some places it is so common as to suggest that most of the present population descends from a single female who was the ancestor of almost all the inhabitants. It is shared with the populations of Taiwan and the Japanese, and shows that the Polynesians spread across the Pacific from Asia and not from Australia. Australian aboriginals and the highlanders of Papua New Guinea do not have this genetic signature. They descend from a migration which began thousands of years before that of the Polynesian *arrivistes*.

One thing is clear: the inhabitants of the Pacific and those of South America have few genetic links. Thor Heyerdahl's book of his intrepid voyage in a balsa raft across eight thousand miles of Pacific from Peru has sold twenty million copies, more than all other anthropology books put together. Unfortunately, his view that to reconstruct the past it is necessary only to relive it is wrong. Population genetics has sunk the Kon-Tiki.

Twenty thousand years ago, much of the Pacific had a

dense population and a prosperous economy. In Europe, too, trade was well advanced. Flint for stone tools was transported for many miles and Baltic amber reached the Mediterranean. There was a brief rise of art, perhaps a mere couple of centuries long, which filled the caves at Lascaux and Altamira with images.

While the world economy boomed the Americas were empty. They were at last reached from Siberia. Many of the inhabitants of that icy land, which was even colder than it is today, lived by hunting mammoths. As they spread they destroyed their food sources. At last, they came to the Bering Land Bridge which joined Asia to Alaska. It emerged from the sea, as did thousands of square miles of coastal plains all over the world, as water was locked into the ice. At the end of the ice age the water rose and twelve thousand years ago the bridge between Old and New Worlds was breached. Just before it disappeared, a few pioneers made their way across. If their experiences were like those of what we know of the nineteenth-century Inuit who made long voyages across such barren landscapes they had a grim time. Many must have starved. Nevertheless, some reached the broad plains of North America and soon spread to the continent's southern point, reaching it within a couple of thousand years. This seems like a rapid expansion but is, after all, less than ten miles a year into a deserted land. The journey was helped by a brief warming which meant that, even in Alaska, a few trees appeared in the bitter landscape.

Once again, the edible inhabitants suffered. Mammoths, sloths, giant tapirs and camels followed each other into extinction. Each was large, tasty, naive and tame. They reproduced slowly. Once humans had arrived their fate was certain. The wave of destruction tempted the first Americans south until, in Patagonia, they could go no further.

The date of the American invasion is not certain. The oldest traces of occupation in North America are in a rock shelter in Pennsylvania. They date from about twelve thousand years ago. Soon, members of the 'Clovis culture', in what is now the United States, could produce sharp and effective arrowheads. The first art in the Americas is at the cave of Pedra Furada – the Perforated Rock – in Brazil, which has twelve-thousand year-old images of birds, deer and armadillos, together with human stick figures. Some claim that charcoal from nearby caves dates back for fifty thousand years, but few anthropologists accept this as evidence of human occupation. Most believe that the first Americans arrived less than twenty-five thousand years before the present.

The genes of Native Americans fit the idea of a small band that filled a new-found land. Americans as a whole are less diverse and more uniform than are the peoples of highland Papua New Guinea (who fill a tiny proportion of the space). The mitochondrial genes of all Native Americans fall into four major lineages as a hint that just a small group managed to complete the hazardous traverse of the Bering Bridge. The same ones are found in some three-thousand year-old Chilean mummies, implying that there were not many bottlenecks on the way through the Americas from north to south. The mitochondria of South American Indians resemble those of north-east Asia, supporting the idea that their ancestors, like those of Polynesians, came from that part of the world (although there is a hint of an ancient link with Europe in a few Northern tribes, suggesting that a more distant traveller across the land-bridge helped found some American groups).

By ten thousand years before the present, humans had filled the whole habitable world, apart from some remote islands. Everywhere they lived in small bands. Every Englishman needed ten square miles of land to feed himself.

The global spread was accompanied by technical advances in axes, arrowheads and nets as the animals easiest to exploit – reindeer, mammoths, giant kangaroos or emus – disappeared and the hunters were forced to move to less easy prey.

The genes of the few modern peoples who still live as hunters and gatherers are a window into that way of life. Adjacent groups often differ quite markedly from each other, evidence that their social structure led to genetic isolation. There were more opportunities for random change as each band split and moved on as the globe was filled. No doubt the days of a hunter-gatherer were rather lonely. Although the immediate group may have been close-knit, there was little contact with anyone else.

Eight thousand years ago, everything changed. There was an economic breakthrough that was to shape the society and the genes of the modern world. Farming began.

Before agriculture, people ate dozens of kinds of food. An excavation in Syria uncovered more than a hundred and fifty kinds of edible plant, but after the onset of farming the diet shrank, to a few cereals and pulses. Even in the nineteenth century, Queensland aborigines ate two hundred and forty different species of plant. To add together the top five crops in the world today gives a global total of just a hundred and thirty kinds.

Hunters had an easier time than did the first farmers. The few !Kung Bushmen who until recently lived in this way needed to work for just fifteen hours a week to feed their families, far less than those who moved to the farming economy (and less than the time which most Europeans have to spend at work to pay the weekly food bill). In the Middle East, too, wild grasses are abundant enough to allow a family armed with primitive sickles to gather enough seeds in a few weeks to feed themselves for a year. Perhaps the extra effort explains the Bible's disparaging tone about the new economic system: Adam, on the expulsion from his hunter-gathering Eden was admonished: 'Cursed is the ground because of you; in toil you shall eat of it all your life . . . therefore the Lord God sent him forth from the Garden of Eden to till the ground from which he was taken.'

The earliest farmers lived in the Middle East, most of them around the headwaters of the Tigris and Euphrates, in a tight core of fertile land in what is now south-Eastern Turkey and northern Syria. Later, farming appeared in the basin of the River Jordan (which is close to where the Biblical Eden must have been). There was plenty of natural food around in what was then a fairly verdant landscape. It was difficult to move elsewhere when times got bad, because of the deserts all around. Rather less than ten millennia ago the weather began to change. There had been a continental climate rather like that of the Midwest of the United States today. Winters were cold and wet and the summer was hot with plenty of rain. Suddenly it shifted towards a Mediterranean climate with warm wet winters and hot dry summers. The lake of Jordan itself began to dry up, and its fresh waters split into the salty Sea of Galilee and the Dead Sea.

Pollen shows that the plants began to change too. The forests shrank and grasses took over. Mediterranean climates are good at fostering the evolution of new plants. Soon there were new and fertile hybrids between grass species that came together as the countryside dried. The local people burnt the grass to attract deer and gazelles to its new shoots. In a few years, they began to plant the seeds, and farming began. Einkorn wheat – one of the ancestors of today's crops – was domesticated close to the Tigris, the relative with which it hybridised in a great crescent from today's Iraq to Israel. Barley, lentils, peas and bitter vetch all found their home within a few scores of miles nearby. Farming itself may have been a very local pastime for a thousand years and more, before the crops and their guardians began to fill the Fertile Crescent about seven thousand years before the present. The teeth of those ancient agriculturalists are worn, because the first grains were milled on soft grindstones and their food was full of grit.

The same sort of thing happened at about the same time in other places. After a transition period in which grass was burned to harvest the new shoots or wild stands of vegetation were watered, agriculture spread at a great rate. Wheat was first cultivated in the Middle East, rice in China and maize in South America. Somewhat later came the domestication of sorghum, millet and yams in West Africa. The effect was always the same: a population explosion. Before farming, each person needed about a square mile to feed himself. After it, a hundred people could live off the same space.

Fossil bones suggest that the health of farmers, far from improving, got worse. Deficiency diseases appeared as the amount of protein went down and there were periods of starvation as population outgrew resources. If children eat well, they grow up tall. This is why the average height in most Western countries has gone up by three inches in the past century. For the children of the first farmers – like those of the proletariat of the Industrial Revolution – the opposite happened. In south-east Europe the average height of men fell by seven inches in the millennium when farming began. The bones of North Americans show extensive damage, most of all in the eye sockets, as maize became the main foodstuff. Maize has little iron and, even worse, reduces the absorption of that essential mineral from other sources such as meat. This led to an outbreak of anaemia, whose record is preserved in the skulls of those who depended on the new maize economy.

Population growth meant that the new habits soon spread. Waves of technical change radiated from each centre of origin. In Europe, decorated beakers appear in archaeological digs, and in the Far East implements of rice cultivation spread for thousands of miles from their Chinese homeland.

From its origin in the Middle East about ten thousand years ago, agriculture reached Greece about five thousand BC and took more than two millennia to cross Europe. Its expansion was not regular. The frontier was rather like that of the nineteenth-century Wild West. The colonists settled the best areas first and left the less valuable lands to their original inhabitants. In north and east Europe, hunter-gatherers managed to stall the wave of farmers from the Danube basin for a thousand years. Their northward spread was further slowed by a worsening climate which made it hard to grow crops. The new technology did not reach Britain until about five thousand years ago. Elsewhere, it was delayed for even longer and in southern Finland the novel economy did not begin until after the time of Christ.

Much of the resistance to the farming way of life was due to the success of the hunters of the 'Forest Neolithic'. Nine thousand years ago northern Europe had a population of affluent foragers. They lived in large camps, built traps for their prey, and stored great caches of food. Around the Baltic, they built stilt villages in ice-dammed lakes. In some places, hunters specialised on seals and in others on deer. Those who gathered ate thirty or more different plants – grasses, acorns, sorrel and dandelions and, in marshy places, water-chestnuts. Millions of broken water-chestnut shells have been found, together with the wooden mallets used to smash them. The single crop was flax, used for rope rather than food.

Wherever farming arrived, the local hunter-gatherers

suffered, sooner or later, a process of gentrification as a wave of economically advanced people moved in on them. It is easy to imagine the complaints of the natives as the newcomers with their new-fangled ways and high technology disrupted their rural idyll. Life in southern England five thousand years ago had quite a lot in common with that depicted in the BBC radio series *The Archers* today.

The farmers may have overwhelmed the hunters but there was a long period of coexistence. Farmers traded grain for meat and furs. In some places, the transition from the old to the new economy took a thousand years, with a slow decline in the number of bones of wild pigs and deer and of natural grasses (as shown by the impressions of their seeds in fragments of pottery) in favour of cattle and grains. The decline in climate at last put paid to hunting. Oysters and seals disappeared from the Baltic and the northern hunters at last moved into the modem world.

Economic historians have two views of the origin of technology. One theory has it that knowledge itself moves, rather than the people who know: new methods are passed from group to group. The other claims that cultural advance comes from displacement and the conquest of one people by another. The sophisticated bring their knowledge with them and replace their predecessors. Bones, pots and seeds hint at the nature of the European Community ten thousand years ago; but the genes say more. Genetic patterns in today's Europeans show that both migration and diffusion were involved in the replacement of hunting with agriculture. The farmers did move in on the hunters, but, just as in *The Archers*, social barriers did not stop sex across the class divide.

To reconstruct the history of Europe from genes is difficult, because it is one of the more tedious parts of the world, with rather little change from place to place. A genetic map based on dozens of genes from hundreds of places does hint at a general trend from south-east to north-west, from Greece to Ireland. This map looks rather like that of the wave of advance of farming, based on the spread of agricultural implements. Farmers moved on at about a kilometre per year by founding new farms at the edge of their expanding population. They interbred with the local hunters and, because they were much more numerous, absorbed their genes. This process began in the Balkans and was completed thousands of years later on the western fringes of Europe. By the time the farmers reached the far north and west their genes had been much diluted with those of the aboriginal Europeans. As a result, the British contain more hunting DNA than do, say, the Greeks, who descend from a less adulterated wave of immigrants who had rolled over the earlier economy and absorbed its genes. The biological heritage of hunters and farmers means that today's Britons are more related to the Portuguese than to the Serbs. The latter live about the same distance away but are closer to the Middle Eastern source of agriculture.

The history of European women – as defined by mitochondrial genes – is not quite so clear. They are divided into half a dozen or so major clusters and a few minor groups, most of whom split apart at some time during the end of the Stone Age, perhaps as the ice retreated across the continent. Although the frequency of each varies from place to place, the geography of mitochondria is harder to fit into a historical narrative than is that of genes that pass through both sexes. There is a hint of an east-west trend along the Mediterranean coast, which might reflect the movement of farmers' wives, but no sign of this further north.

The genetic map of Europe has a few anomalies. The Basques do not fit into the general pattern. They have a number of unique features, with the highest frequency of the Rhesus negative blood group gene in the world. Excavations show that the locals resisted the new technology for thousands of years. They still differ from all other Europeans and may be closer to our hunting ancestors than anyone else (although their mitochondria are not distinct from those of their neighbours). The Lapps, too, are distinct and descend from a different group of hunters, whose society they still in part retain. Sardinians are also rather different from the rest of Europe and have affinities with the Basques. The remoteness of their island home may have reduced the number of immigrants.

There are also genetic trends away from a Middle-Eastern centre to the north-east towards Siberia, the south-east in the direction of India, and, more ambiguously, south-west into North Africa. Perhaps these too reflect of a wave of farmers on the move away from a population explosion who absorbed the genes of the local inhabitants as they spread. One North African group, the Berbers, now scattered in tribes across Morocco, Algeria, Tunisia, Libya, and Egypt is shown by its mitochondria to be distinct from the Arabic-speaking peoples who surround them. They fit instead into the European family of female lineages and may be the remnants of another branch of the first wave of farmers, who passed south of the Mediterranean.

The new economy left analogous trails in other parts of the world. Rice cultivation started in the Yangtse basin about eight thousand years ago. Within three thousand years there were rice farmers from Vietnam to Thailand and north India. These were the people who developed sea-going canoes and spread into the remote Pacific, where – because rice cannot be grown there – they planted breadfruit, taro and yams. The pollen record from three millennia ago shows that large parts of Java were intensively farmed. Because they were entering an empty land the genes of these Pacific farmers and fishermen are still quite similar to those of their Asian ancestors. In Africa, too, there was a population explosion in places where millet was first grown. The movements of the sickle-cell gene can be traced across the continent in the wake of the first farmers. Those people were in their own way as destructive as had been their hunting ancestors. Great and empty cities in North African deserts were once supported by fields that have now been overwhelmed. In the same way, in Spain, the Mesta, the great cooperative of the shepherds, turned most of the country into a desert within three hundred years.

Those African farmers and their European counterparts no doubt experienced social unrest as they gave up hunting to move to a more productive but perhaps less enjoyable way of life. However, any romantic view of a harmonious past when contented foragers shared their food is a hunger for a nonexistent Golden Age. Virgil, in the *Georgics*, mourns for a time when 'No fences parted fields, nor marks nor bounds,/ Divided acres of litigious grounds.' His plaints over a happier past may have been shared by the early farmers as they mourned the glorious times when they hunted food rather than growing it. Whatever the truth, the origin of agriculture marked the end of an economic system based on individual effort which lasted for nine tenths of history. With farming, Eden had been left forever; and politics began.

Chapter Eleven

THE KINGDOMS OF CAIN

Adam and Eve's children were a worry to their parents. Their eldest, Cain, is best known for having killed his brother Abel. He has another distinction. Just one generation after the expulsion from Eden he became the first capitalist. As the Old Testament says, he was the earliest to 'set bounds to fields'. By so doing he erected barriers among the peoples of the world. Frontiers have driven society, history, and genes ever since.

No doubt the idea which came to Cain struck the first farmers as well. Ownership of land was born with agriculture. The process can be seen today as hunter-gatherers give up the old social order. The Kipsigis of Kenya moved to a settled existence as maize farmers in the first years of the twentieth century. Great inequalities soon appeared. When harvests were bad the poor starved while the rich grew fat. Competition among males to gain a mate increased and there was a new campaign in the battle of the sexes. Those who owned productive land had far more children than did those with none. The farming genesis was when class began. From Mycenae to ancient Chile there emerged a difference in height and health between the rich, interred with their ornaments, and the rural poor, buried in penury.

The first farmers soon argued about who was to grow what and where. It did not take long for property to pass into fewer hands and for society to evolve into the system of competing tribes that persists today. Any barrier, be it a mountain, a frontier, or an inability to understand, which stops peoples from meeting and mating will cause them to diverge. All over the world, genetic changes mark the divisions – the bounds to fields – between ancient societies.

Even so, politics is a new element in the evolutionary equation. Genetics suggests that what we see as history, the struggles between nations, is a recent event. From the Old Testament to *Mein Kampf*, historians have seen conquest as the key to the peoples of the world. In the turbulent years after the First World War, the League of Nations tried to define just what a 'nation' might be. The best they could come up with was 'a society possessing the means of making war'. Over the past millennium, most great nations have spent half their time at war; but marauding states have shaped biological history only in the past few thousand years. Before then, people and their genes moved by gradual diffusion or by migration into an empty land, rather than by the defeat of one social entity by another.

In many parts of the world the earliest farms, and the first settled societies, were by rivers in an arid landscape. Such rivers (the Nile most of all) often flood, to leave fertile silt as they recede. Modern tribal farmers who use the land left bare by the departing waters of the Senegal River obtain a return on labour of fifteen thousand per cent: for every calorie of effort they put in they get a hundred and fifty back as food. This compares with a return of around fifty to one for the most efficient modern fields.

The return on the flood plain is enormous but, like the stock market, unpredictable. For the locals, life, as on Wall Street, could be bumpy. The flooding of the Nile has been noted since AD641. The records reveal a hundredfold difference in the area of land submerged from year to year. Some years are excellent but others are dry and disastrous. In today's Senegal, with its equally capricious rivers, this has produced a rigid pecking order. Some families always have access to the floodlands even when the area inundated is small. Others are allowed to grow crops only when the river has risen high and covered large tracts of ground. In dry years they have to find food elsewhere, and in earlier times that meant a return to hunting. Perhaps the earliest settled communities developed, not to increase efficiency but to manage risk. A wild free-for-all for the best land in a bad year would have been dangerous and expensive. Society evolved as a way of coping with uncertainty.

Ten thousand years ago the Natufians, the descendants of the cultivators of the Jordan Valley, had built villages with timber houses. Within two millennia Mesopotamia contained much larger settlements. It took but a few centuries for civilisation to advance to such an extent that such places were surrounded by walls, ditches and watchtowers. Warfare had begun to play the part which it has retained ever since. Farmers were forced from their hamlets by land degradation and the pressure of numbers. In Mesopotamia they moved into the hot and dry plains away from the Tigris and Euphrates rivers. Soon, the earliest city-states began, perhaps because of the need to organise which began with the irrigation. For the first time humanity was divided by political rather than physical barriers. The genes of today show that, since then, bigotry has been as effective an obstacle as has geography.

Capitalism was helped by technology. The bones of six-thousand-year-old horses at Sredny Stog in the Ukraine have broken teeth, as if they were controlled by bits. A horse increases mobility and helps people to work together to steal from others. Its power is seen in the success of a few dozen Spaniards in the conquest of the Inca and Aztec Empires and of the Mongols in taking over Hungary. Soon after the appearance of horsemen the civilisations of Eastern Europe built defensive walls around their towns. Within a few years their societies had collapsed.

By 3600 BC Mesopotamia contained great cities. Uruk had ten thousand people and within a millennium that number had increased fivefold. Its growth was due in part to warfare. Scores of villages were abandoned by their people, who moved to the new cities. The Sumerian city states, the first organised political entities, were the source of writing and of wheeled transport. They had a priesthood and an aristocratic caste, and a dispossessed mass. Their decline was hastened by mismanagement. With irrigation, the soil became salty and in the last years of Sumeria crop production dropped to a third of its peak. These, the first nations, were overcome by one of the first empires, that of the Akkadians, who invaded from the north.

Other cities came to an end because of bad planning. The ruins of Petra, in Jordan, are today surrounded by miles of arid desert. The evidence of its decline is preserved in an unusual way. Hyraxes (small mammals about the size of a guinea pig) live in communal mounds. They have the singular habit of cementing their homes together with urine, which dries to form an unpleasant but effective glue. It also preserves the seeds upon which their ancestors fed. At its height, Petra was surrounded by forests of cedar and pine. These were burned. Grassland followed and this was much farmed. Within a few centuries, the desert had taken over. No doubt, the inhabitants of Petra in its last days fled the city, taking their genes with them.

No one has studied the patterns of genes in today's Iraqis or Jordanians (some of whom may be direct descendants of the people of Petra). When they do, the genetic relics of the first cities may be revealed. However, other forgotten societies – and the divisions between them – have left biological traces which persist to the present time.

Soon after the collapse of the Sumerians, the Greek *polis* or city-state appeared. Its philosophy – and its name – is at the basis of modern politics. The *Iliad* and the *Odyssey* are accounts of wars among the first *poleis*, which included Corinth, Sparta and Athens. Greece entered its classical age. This was a triumph in artistic, economic and political terms. Three thousand years ago Greece was the most densely populated country in Europe. Its inventive people expanded to form Greater Greece, Magna Graeca, an empire that extended from the Caucasus to Spain. Forty towns in southern Italy were Greek. They included Syracuse, then the biggest city in the world, and Sybaris, a byword for wealth.

Patterns of blood groups and enzymes show that today's southern Italians and Sicilians are distinct from their compatriots to the north and share many genes with the population of modern Greece. The DNA of the first European states remains as a witness to their past. Sardinians, too, owe their distinctiveness to an ancient nation-state. They are related to the modern Lebanese, whose country occupies the territory of the Phoenicians, once the greatest traders of the Mediterranean.

Greeks, unlike Sumerians or Phoenicians, are still around as a reminder of the past. At about the time of the Greek Empire, central Italy supported another buoyant economy; that of the Etruscans, now the embodiment of obscurity. They lived in cities of half a million people and were skilled metal workers with, according to their Latin neighbours, a feminine and dreamy personality. Dreamy though they were, for a brief period their empire encompassed Rome itself. Almost no relics are left. The word 'Tuscany' refers to their homeland, and some enigmatic sculptures with a characteristic smile remain as does an eccentric object, a bronze sheep's liver covered in messages, used as a crib by the priest as he disembowelled the sacrificial lamb. These hints from the past were, until recently, all we knew about the Etruscan nation.

Its heritage has not been lost. Between the Rivers Arno

and Tiber – modern Umbria – is a region distinct from its neighbours. It retains some of the genes of the Etruscans. Their biological legacy lives on in their descendants, although their language and culture are long gone.

Many movements involved commerce rather than conquest. Often, the traders left genetical calling-cards. The Silk Road passes from the ancient Chinese city of Changan to the Mediterranean. It has been a trade route for more than two thousand years and for much of this period was the main artery of cultural exchange. Silk passed from east to west; in return came cotton, pomegranates and Buddhism. Modern China has few of the genetic variants in haemoglobin, the red blood pigment, which are common elsewhere in the world, but the blood of today's Silk Road reveals a trail of variant haemoglobin genes. They came from the Mediterranean and spread, with the traders, along this ancient trackway. At its western end in China, about one person in two hundred carries an abnormal haemoglobin, while at the distant eastern end this drops to one in a thousand.

Other dispersals involved forced migration. Stalin moved thousands of people from the Crimea to Central Asia, and today's movements of minorities across Eastern Europe after the collapse of Communist regimes will have genetical consequences (although in the Balkans at least it seems that past turmoils have already led to so much blending that ethnic boundaries, cause of conflict as they are, do not reflect genetic change).

Often, there have been attempts to re-unite peoples fragmented by history. In the 1920s, Greeks (whose ancestry could be traced from Magna Graeca itself) were exchanged with Turks who found themselves marooned in modern Greece. Greek-speakers, many of whom incorporated Byzantine genes, were moved from as far east as the Caucasus. Part of the drive toward nationhood was the desire for unity among peoples who share a culture. Dr Johnson put it well: 'Languages are the pedigree of nations'.

The potency of speech in forming an identity is illustrated by the Statutes of Kilkenny of 1367. At that time, the English had subdued only that part of Ireland around Dublin known as the Pale. Everything beyond was seen as barbarous. The authorities were alarmed by the encroachment of the natives, which went as far as marriage with the settlers. The Statutes declared that '. . . now, many English, forsaking the English language, manners, mode of riding, laws and usages, live and govern themselves according to the manner, fashion and language of the Irish enemies, and have also made divers marriages and alliances between themselves and the Irish enemies, whereby the said land and the liege people thereof, the English language, the allegiance due to our Lord the King, and the English laws are put into subjection and decayed, and the Irish enemies exalted and raised up contrary to reason . . . Therefore, if any Englishman or Irishman dwelling among the English, use Irish speech, he shall be attainted and his lands go to his lord.'

In Ireland the Dublin government still struggles to save the almost extinct Irish speech of the Gaeltacht and, in a matching historical obsession north of the border, not until 1992 was the ban on the use of Gaelic street names in Northern Ireland lifted. For six hundred years two nations who share a small island have tried to retain their identity with language; an attempt which, bizarrely enough, has survived the death of one of the tongues involved.

Any entity, be it a language or a pool of genes, which remains isolated from its fellows will begin to evolve away from them. Biological evolution has parallels in the origins of new languages from a shared ancestor. The analogy between linguistic and biological change is a deep one. Language barriers slow the movement of genes, and linguistic obstacles may mark a genetic step. What is more, trees of language sometimes resemble those of genes, as a hint about a common history.

The world has five thousand different languages. Many more – like Etruscan – are extinct. Like genes, languages evolve because they accumulate mutations. Some words change quickly while others are more conservative. Although the Victorians claimed that within a hundred years English and American would be mutually unintelligible, most languages retain enough of their identity for a sufficient long period of time to be, like genes, clues about the past.

Sometimes the barriers are scarcely noticeable. England can be divided into zones defined by whether people do or do not pronounce the final letter 'r' in words such as 'car'. I do not – I say caH because I was brought up in Wales and on Merseyside, but many of those in Cornwall, Lincolnshire or Northumbria (and plenty of Americans) pronounce it as caR. This might seems trivial, but such tiny differences can mount up to build a barrier to the exchange of information until a new language – and often a new people – is born.

Italy has several dialects, some of which trace certain words to their Greek past. Other local tongues also reflect history. A Portuguese farmer can no more understand a Venetian than we can, but he can talk to his Spanish neighbour, who can converse with his Catalan cousin, who in turn is linked to Italy through the *langue d'oc* in southern France. The chain of words reflects a history of shared descent that traces back to the Roman Empire and before.

It is sometimes possible to guess at what the ancestral languages sounded like. Father, *padre* and *pére* are obviously related terms. They all descend from the same word, p'ter; which means that the phrase 'God the father' can appear as both *deus patris* and *Jupiter* or, in Sanskrit, as *diu piter*.

A pedigree of European languages shows that nearly all are related. This Indo-European family also includes Indian languages such as Bengali and extinct tongues like Sanskrit. Its existence was recognised by Sir William Jones in 1786, who saw that Greek, Latin and Sanskrit 'all sprang from some common source which perhaps no longer exists'. Finnish, Hungarian, Turkish and Maltese belong to other linguistic groups, but half the world's population now speaks an Indo-European tongue.

A political map of modern Europe is crossed by many national barriers, most of which mark a shift in language. Most Frenchmen speak French, and most Germans German. Language is a force for national cohesion and a barrier to the movement of people. It reduces the chance of marriage and the spread of genes and has long done so. The Old Testament describes the fate of an Ephraimite prisoner taken by the Gileadites: 'The men of Gilead said unto him, Art thou an Ephraimite? If he said nay, then said they unto him, Say now Shibboleth; and he said Sibolleth: for he could not frame to pronounce it right. Then they took him and slew him.'

Some borders between languages are regions of biological change. There are genetic differences between Welsh and English speakers in part of Pembrokeshire. This 'Little England beyond Wales' began when in 1108 King Henry I moved a group of artisans to Wales from the banks of the Tweed to set up a weaving industry. Their anglophone home ended at a sharp boundary, the Landsker. Even a century ago, just one marriage in five hundred took place across the divide. Eight hundred years after they arrived, the blood groups of the descendants of the immigrants still differ from those of their Welsh-speaking neighbours.

In the same way, the population of Orkney, whose native language is a Scandinavian one, is distinct from the rest of Scotland. Even dialects may mark biological barriers. France has a small genetic step between those who speak the *langue d'oc* (southern French) and speakers of the northern *langue d'oïl*. Genes and language tell the same story.

Their concordance is not always absolute. The Balkans have had, and retain, a tumultuous history of movement and conquest which has obscured any relationship between linguistic and genetic units. Hungarians, too, speak a distinct language, but remain biologically close to their neighbours. The Magyar conquerors from the east imposed a language on their subjects, but were too few in number to make much impact on the genes. In some places genetic steps exist even within groups who speak the same language. The east of Iceland is somewhat distinct in its genes from the west, although both speak Icelandic. This may be a relic of the history of settlement of western Iceland by Scandinavian immigrants who brought wives and servants from Ireland.

The language of the Basques, like their genes, seems to be unrelated to any other. The Latin author Mela wrote of his bafflement at the names of peoples and rivers which meant nothing in any tongue known to him. Francis Galton himself, who often went on holiday to the Basque country, recalled 'the legend that Satan himself came here for a visit. Finding after six years that he could neither learn the language nor make the Basques understand his, he left the country in despair.' Satan's problem is illustrated by a typically impenetrable Basque proverb: '*Oinak zewrbitz-atzen du eskua, eta eskuak oina*' ('the foot serves the hand and the hand serves the foot'). Basque may be the last remnant of the speech of Europe before agriculture. Its sole apparent relative is spoken by some of the isolated peoples of what was once Soviet Georgia. Many Georgians believe that their own language was taken to the Basque country by Tubal, grandson of Noah, and moves were once afoot to find a Basque to succeed to the throne of Georgia.

Safe in their mountains the Basques resisted invaders, so that their ancient tongue, a language of hunter-gatherers, lives on among its half-million speakers. The skeleton of Cro-Magnon himself was found in a part of France which, its place-names suggest, was once in the Basque country. There might even be a linguistic, and perhaps a genetic, link between Cro-Magnon, one of the first Europeans, and the Basques. This last remnant of a lost European economy is under threat. Today, Basque genes stretch for much further than the language: east to Zaragoza, now a Spanish city, and north into France. Their economy was destroyed long ago. Now their speech and their culture may at last be squeezed out by modern society. Like the Etruscans – who also spoke a non-Indo-European language – only their genes will be left

DNA can say much more about ancestry than can language. The forefathers of today's Britons come from Europe, Africa, India and even China, but they speak English. Books and pictures will in time blur the links between genes and language, but we are still in a phase of history where enough remains of the linguistic past to speculate about the origins of speech and perhaps of modern man himself.

Where did Indo-European languages come from? The first recognisable member of the group was Hittite, written in cuneiform and spoken in Turkey four thousand years ago. Modern Indo-European tongues can sound quite different. 'Our father, who art in heaven' is *'Ein Tad, yr hwn wyt yn y nefoedd'* in Welsh, *'Patera mas, pou eisai stous ouranous'* in Greek, *'Otche nash, suscij na nebesach'*

The Kingdoms of Cain

in Russian and '*He hamare svargbast pita*' in Hindi.

Nevertheless, some words for widely used objects are held in common. They can be used to guess where the languages originated. Indo-European tongues share several terms for domestic animals and crops. The ancient word for sheep, *owis*, has been inferred from the Latin *ovis*, Sanskrit *avis* and English ewe. Cow was *kou*, and water *yotor*. There are similar words for corn, yoke, horse, and wheel, too.

Perhaps the Indo-Europeans were farmers, who brought their language with them as they spread. Quite where their homeland may have been is uncertain. Their language began long before the first record was preserved. They might represent a wave of invasion of Kurgan people from the Pontic steppes, north and east of the Black Sea. This was the land of the Sredny Stog horsemen. Their journeys began about 4500 BC, long after the origin of agriculture. Some believe that the Indo-Europeans invaded much earlier and brought farming with them as they migrated from Asia Minor three thousand years before the Kurgans. Some of the Indo-European peoples – and languages – who built Europe may have begun to diverge before they moved from their homes in the east. It may be hard to trace just who, if anyone, among today's nations and tongues are the ancestors of modern Europeans.

Language, archaeology and genes all bear witness to an invasion of Europe from the east. Farming, genes and speech are intimately related. In some place, farmers moved into an empty – or scarcely populated – land rather than into a successful hunting economy (as it did in Europe). Rice growers of the Far East around the Yangtse basin took their language as well as their DNA with them as they filled the Pacific. At one time these Austronesian languages were the most widespread of all, with a territory that extended from Madagascar to Hawaii and Easter Island.

In Africa, farmers moved south, to fill western and southern parts with Bantu speakers.

Today's technical advances, from the book to the internet, provide new ways of speaking to people. They have led to the erosion of the nation-states that have so long shaped history. We can now reach anyone in the world as soon as they can get to a telephone. New work on global patterns of language suggests that the first social breakthrough of all also involved a new form of communication technology.

The patterns of genetic change which build up through mutation can be used to make an international pedigree. Africans form a distinct and ancient branch of the lineage. American Indians group together with their Asiatic ancestors, and Australia and New Guinea are a separate offshoot. A family tree of languages can be made in the same way. English, German and Bengali cluster together into the Indo-European family, and Chinese and Japanese into a different group. A language tree based on a few words – one, two and three; head, ear and eye; nose, mouth and tooth and so on – looks much like one made with a more complete vocabulary. Such limited word lists are used to classify less well known tongues (such as those of Africa or the New World) with some success.

One controversial claim has it that all the languages of the world can be classified into just seventeen distinct families, with the thousand or so native languages of the Americas failing into only three; Eskimo-Aleut in the far north, Na-Dene in southern Alaska and Canada, and all the others south to Patagonia as a single group, Amerindian. The wide distribution of this family contrasts with the pattern in Papua New Guinea, where a much smaller space contains eight hundred languages, many almost unrelated to each other. The genetics and the speech of the Americas and of Papua New Guinea shows parallel patterns: Ameri-

The Kingdoms of Cain

cans are rather uniform in their DNA and in their language, while the Papuans vary from valley to valley. The highest concentration of language diversity lies in the Caucasus, between the Black and Caspian Seas. In an area twice that of Britain forty languages are spoken, some in just a single village. Unfortunately we know little of the genetics of that fractious part of the world.

A tree of the relationships of all the world's languages makes it possible to guess – wildly – at some of the original words at the base of them all. Russian linguists have attempted to reinvent Nostratic, the twelve-thousand-year-old tongue thought to be the ancestor of Indo-European and its relatives. These include the Elemo-Dravidian tongues of parts of India, the Altaic languages which include Turkic and Mongolian and an Afro-Asiatic group spoken in the northern half of Africa. They have reconstructed over a thousand 'root' words. *Tik*, for digit, finger or toe, is one of these, *kujna* for dog another. No shared words refer to agriculture, so that this proto-language may indeed derive from before the farmers.

The world language tree looks somewhat similar to the genetic tree. Both come to the same root in Africa and both show a split between Australasia and other Asian peoples. Not too much should be made of this, as words can spread by learning, which genes cannot. As a result, the pedigree of words looks more like a network than a branching river (to eat an avocado while paddling a kayak unites three distant families in a single sentence). In addition, trees of genes need not always reflect that of the population from which they come (particularly if certain genes spread through whole groups because – as for malaria resistance – they are advantageous). Even so, the general similarity between the two means of communication suggests that perhaps language itself dates back to the origin of humankind.

Speech marks a huge jump in the speed of information transfer. To spell out this sentence, letter by letter, would take ten times longer to transmit the information than it would to speak it. The plight of the deaf and dumb shows how much of life depends on an ability to speak. Dyslexia – a difficulty in recognising written words, often among people otherwise of high intelligence – has been tracked down in part to specific genes on two human chromosomes; and these may be candidates for that select group that may differentiate ourselves from our primate relatives. It is hard to imagine a society which could work without language. Early modern humans underwent changes in skull shape and in the position of the larynx that may have marked the first ability to articulate a sound. Such physical changes suggest that speech may have made us human in the first place.

Shelley felt as much: in *Prometheus Unbound* he has his hero 'give men speech, and speech created thought'. Not everyone agrees. Some suggest that even Neanderthals had a sophisticated language which disappeared when they themselves became extinct. There is a hint of an earlier linguistic dawn. Apes in groups spend much of their time grooming, to show their fellows that they belong. If the first humans reassured their companions as apes do, they might, because of the size of each band, have had to spend half their time grooming. Speech, even when primitive, is a better way of calming one's fellows than is touch. The first sentences may have been words of comfort.

Nobody will ever be able to speak Neanderthalish, if it existed. The sixteenth-century German philosopher Becanus was convinced that the language of Eden was Old German, and that the Old Testament had been translated from this into Hebrew (the Emperor Charles V, in contrast, spoke French to men, Italian to women, Spanish to God and German to horses). Soon, there may be a chance to find out the truth. The fossils and the genes have already given us clues about where and when Adam met Eve; before long, we may be able to guess at what they said to their errant children.

Chapter Twelve

DARWIN'S STRATEGIST

American bird-watchers know that the common sparrow – the bird that hops around in English gardens – has a bigger body and shorter legs in the north than in the south of the United States. The same is true for sparrows in northern and southern Europe. Creationists see in this a divine arrangement to ensure so that each species fits into the economy of nature; cold places, wherever they are, meriting a subtle change in God's plan.

If the deity does have a plan, it seems to work in the same way for humans. People from the far north have shorter arms and legs and more compact bodies than do those from the tropics. Olympic long-distance records tumbled after East Africans with their long legs began to take part. Before Darwin the ability of Africans to cope with heat and Eskimos with cold was excellent evidence for divine action. The Creator had seen to it that each people suited their homeland, as proof of what a wonderful designer he was. As the nineteenth century cleric William Paley argued, if one found a watch, beautifully designed as it was, then one must accept the existence of a watch-maker. The perfection of humanity proved in the same way that there was a God. This idea seemed so powerful that it was carried to absurd lengths. Voltaire, in *Candide*, parodied it with Dr Pangloss and his delight at the perfection with which noses had been designed to carry spectacles. Freud, a keen Darwinist, commented that one might just as well argue that the fact that cats have two holes in their skin where their eyes are could be explained in the same way.

The argument from design, as it is called, has a problem, for sparrows at least. In fact, English sparrows have not been in the Americas since the time of creation. They arrived little more than a hundred years ago. A few were brought from England and released in Brooklyn in the 1850s. Within about a century, a hundred sparrow generations, they spread to fill the continent. How did they come to resemble so closely the birds of their native land?

The answer lies in natural selection: in inherited differences in survival and reproduction. Studies of marked sparrows in Kansas show that large individuals with short legs survive better in icy weather. They hence have a greater chance to breed and to pass on their genes when spring comes. Those released a century ago brought from their native land genes for large or small size and stocky or graceful legs. In the north, the big squat birds did better, but in those that spread to the torrid south the opposite was true. In a few generations, American sparrows evolved just the same geographic patterns as those found on the other side of the Atlantic. Natural selection had done its work.

Natural selection was Darwin's Big Idea. It gave him a mechanism that drove evolution without the need for a designer to supervise every step. *The Origin of Species* starts with life on the farm. It shows how domestic animals emerged from wild ancestors because of preferences, often inadvertent, for one type over another. Selective breeding, the choice of the best to produce the next generation, soon caused new forms to appear.

If farmers could do so much in a short time, then nature could do more. 'What limit can be put to this power, acting during long ages and rigidly scrutinising the whole constitution, structure and habits of each creature –

favouring the good and rejecting the bad? I can see no limit to this power, in slowly and beautifully adapting each form to the complex relations of life.'

The engine – if not the engineer – of evolutionary change is the preservation of favoured types in the struggle for life. Change is inevitable in any system, be it genes or language, which makes errors of transmission from one generation to the next. This may be evolution, but it is change at random. It cannot lead to the passage from simple to complicated which made humans from their modest predecessors. Natural selection takes advantage of the fact that, each generation, inheritance makes mistakes. Because some improve the ability of their carriers to cope with what nature throws them they copy themselves more successfully. Darwin's mechanism sorts out the best from what accident supplies. It gives a direction to evolution and allows life to escape from the inevitability of extinction. This is as true for humans as for any other creature.

Selection is a simple idea. The notion is used by computer experts. They programme their toy not with the precise details of what is needed, but with a guess of what might work; and allow this to make rough copies of itself. By choosing the most successful, they make rapid progress and can, in a few generations, evolve computer birds that flock like starlings, mathematical ants able to follow trails and programmed flowers as beautiful and unexpected as any product of nature. Literature, too, is not immune to artificial natural selection. Just a few unpretentious themes underlie simple works like mediaeval folk tales or children's stories. To feed them into a computer with a small change to each motif and to choose the best leads to the emergence of brand new and coherent versions.

Humans are not safe from the Darwinian machinery. For most of history, most people died before they were old enough to pass on their genes. Even among the survivors, some had more children and some fewer. If any of these differences are influenced by inheritance, then Darwin's mechanism is at work and the next generation will differ from its parents. Selection will, in time, lead to change.

The power of the evolutionary machine rests on its ability to choose the best available, even if it is not much better than what went before. Lewis Carroll saw how it works. Imagine that we have a three-letter word – 'pig' for example – and we want to change it into another – 'sty'. We can change any letter into any other. If we make random changes and just hope for the best, taking any meaningless set of letters each time, it takes thousands of moves to get the pig into the sty. Natural selection imposes a rule: all the words in between must make sense. It picks up combinations that look good and builds on them. It can get there in just six steps – pig, wig, wag, way, say, sty.

The theory of evolution caused a sensation in 1859, the year of publication of *The Origin*, because it seemed to remove the need for a direct link between god and man. The wife of the Bishop of Worcester said of it: 'Let us hope that it is not true – but, if it is, let us pray that it does not become generally known!' After the establishment had recovered from the shock, religious thinkers came up with the idea that evolution was a means to work out God's plan. Even if humans were not perfect, they were perfectible, and selection was how the deity had chosen to do it. However, its action, far from perfecting the imperfect, often seems incompetent or even cruel. Panglossians can find little comfort here.

Selection can do remarkable things. But much is beyond it. Natural selection cannot plan ahead; it acts, without foresight, taking no thought for the morrow. It does just what is needed and no more, and does it in a slapdash and shortsighted way. It is, to use Richard Dawkins' memorable phrase, a blind watchmaker, achieving an extraordinary end through a simple and inefficient means.

It is easy to claim the whole of biology as evidence of its action. Just as Paley interpreted the complexity of plants and animals as an argument for God, a neo-Paleyism plagues evolutionary theory. It claims that all animal structure is well adapted and must hence always reflect the action of selection. This argument is circular, but is hard to disprove. It has led to many disagreements – fascinating to their proponents, tedious to those outside – among biologists. Some feel that the Darwinian machine drives the whole of evolution from the shape of the nose to the order of bases in the DNA. Others see selection as an occasional event that directs some genes while most change at random. The issue is unresolved and some biologists like to spend their time making up stories about how selection has moulded the most improbable characters. Sometimes they even turn out to be right. Anthropologists have even more vivid imaginations and have made some unlikely guesses about how selection may have formed human attributes. Many are fantasy, but because they invoke events that happened long ago are almost impossible to refute.

Differential survival and reproduction are at work in many places without anybody realising. One egg in thousands and one sperm in millions produce an offspring. Do the rest die at random or do they fail for genetic reasons? Nobody knows: but if only the best survive, the Darwinian machine is a more pervasive force than was ever imagined.

Whatever its importance, selection is just a mechanism and not a force for good. Cancer patients are sometimes given a drug which attacks cells as they divide. The treatment often fails because selection is at work. A few cells have undergone a mutation which changes the properties of a certain gene to enable it to break down the drug.

These soon take over, sometimes so effectively that the patient dies. There is not much evidence of a benign designer here.

Humans show as well as anything else its strengths and weaknesses. *Homo sapiens* has changed as he filled the world over the past 150,000 years. That six thousand human generations is the same as the number of generations of mice that separate today's animals from those that infested the brand new Acropolis and ties modern fruit flies to the insects that swarmed over the apples of William the Conqueror. As far as anyone knows, today's mice and fruit flies have scarcely changed over that time, emphasising just how short a period we have had to evolve.

Three ages of history have moulded natural selection. A lengthy Age of Disaster was followed by a shorter Age of Disease and – very recently – by an Age of Decay. For most of the past nearly all those born died by disaster, through cold, hunger or violence. Many individual tragedies acted as agents of progress. This chapter is about how humanity evolved to cope with the changes in climate and diet as we moved from our African home. The second epoch, the Age of Disease (which began only a few thousand years ago), albeit over in the West, is still the rule elsewhere. The Age of Decay (in which most people die of old age) is now upon us. Because most of those who succumb nowadays have passed on their genes it is hard to know what selection will be able to achieve in this, the third age of human evolution.

Our ancestors, our relatives and ourselves are tropical animals. Noel Coward notwithstanding, humans are among the few large mammals able to cope with the African midday sun. Most people, given the choice, prefer a warm place (perhaps only the Costa del Sol for a couple of weeks a year) and many adaptations have evolved to combat heat, rather than cold. Humans are the least hairy of all primates and sweat the most. On a sunny day, the temperature at, or a few inches above, the ground can be as much as twenty degrees higher than that a couple of feet away from the earth, because the surface absorbs and reflects heat from the sun. Our upright posture may have evolved in response. One of the best ways to reduce heat stress is to stand up, out of the layer of hot air. Perhaps, as our distant ancestors moved to the savannah from the forests, they stood up to cool down; opening, literally and figuratively, new horizons for their descendants.

Humans today live in every environment from rainforest to tundra and from sea-level to five thousand metres above it. Culture – fire, clothes and houses – helped us fill the world, but there have been genetic responses to climate as well.

Mankind left Africa more than a hundred thousand years ago and reached New Zealand, the furthest point of his spread, a thousand years before the present. For much of the time, the weather was even worse than it is today. Ancient climates can be inferred from shifts in the chemical composition of water. In the Arctic, this falls as snow and is preserved as ice. A core has been drilled through three thousand metres of Greenland ice to reach the rock below, where the first snows fell two hundred thousand years ago.

The record of the icecap reveals many ice ages during the evolution of *Homo sapiens*. The last one peaked about eighteen thousand years before the present. It had a drastic effect on the mammals of the world, ourselves included. The giant sloths and native horses went from the Americas, the mammoth from Asia and giant lemurs from Madagascar. Large areas of northern Europe were abandoned. As the climate dried because water was trapped in the ice, parts of Africa became desert and were lost to habitation as well. The level of the sea fell. The Bering Straits and Bass Strait dried out. Broad coastal lowlands emerged in many places. The air filled with dust from the icy deserts. Even if they were cold most of the time, our ancestors had beautiful sunsets.

In the Russian Plain settlements flourished within a hundred and fifty miles of the icecap. The Frenchmen who painted the cave at Lascaux could not relax by basking in the sunshine in a pavement café. The arctic ice was three hundred miles away and they had to stay warm to stay alive. Perhaps the need to keep under cover fostered artistic endeavour. The explosions of artistic and technological style all happened on the northern edges of humanity's range. Humans survived the harsh new climate and at the peak of the last glaciation were the most widely distributed mammal in the world – a status they have retained ever since.

Not all was gloom at the time of the global spread. In brief periods, up to a couple of thousand years long, the temperature rose by as much as seven degrees in just a few decades; a change equivalent to the Scottish climate shifting to that of southern Spain within a lifetime. Perhaps these sudden strange warmings impelled the colonists on their way.

As in the sparrows, differential survival and reproduction favoured those best adapted to climate. The Neanderthals, our extinct cousins, who had lived in a chilly Europe long before the modern upstarts arrived, were short, squat and heavy. Rather like today's Eskimos they were adapted to the cold (so much so, indeed, that the average Neanderthal was more thickset than all but a tenth of modern Eskimos). Most people would change seats if Cro-Magnon, an early European, sat next to them on the tube, but would change trains if a Neanderthal did the same.

Modern humans show geographical trends in body build that reflect the action of climatic selection. Eskimos are about a third heavier for a given height than the world average, while men from parts of East Africa are much slimmer than others, at about three quarters the weight expected for their height. Much of the difference arises from changes in body proportions. Most peoples from the tropics are tall, thin and have long arms and legs. Those from the north tend to be more heavily built. For unknown reasons, the trends are stronger in men than in women. The same is true for the changes in shape in sparrows, perhaps because larger males are more aggressive in the struggle for food in winter. Although little is known about the inheritance of such characters (and environmental effects are without doubt involved) the differences across the globe are at least in part genetic.

The short, fat peoples of the north are better at keeping heat in the body core. Those with more graceful figures from hotter climes cool down better through their long arms and legs. Most of the body's excess heat is lost from the skin and the amount of surface per unit of volume is greater in thin and spindly individuals.

Some populations are even able to regulate the amount of heat that gets to the arms and legs. If a European or an African puts a finger into icy water, its temperature drops to a level low enough to damage the flesh. When an Eskimo does the same his finger stays warm. Again, it is not clear how much the effect is genetic, but among North Atlantic fishermen those of European origin are worse at keeping their hands warm than are the natives of the North. Australian Aborigines have another defence against a climate hot during the day but cold at night. They close down blood vessels near the surface on cold nights, so that their skin temperature falls to well below that of a European in the same conditions, saving heat in the body core. Aborigines are also better able to handle cold without shivering and can sleep in the open without too many problems. Even the rate at which the body uses energy is lower in those who evolved in the tropics.

Other patterns might also be due to climate. The woolly hair of Africans is said to help sweat to evaporate and cool the head down. The long fine noses of peoples from the Middle East might help to moisten the desert air before it reaches the lungs and the narrow eyes of Chinese to protect against the icy winds of the Asian plains. All this is guesswork.

The globe has one anomaly in its climatic trends. In the Old World at least, most tropical peoples have darker skins than do those from cooler climes. As anyone who has sat on an iron park bench on a sunny day knows, black objects heat up more in the sun than do white, so that black skin, far from protecting against the sun's heat, soaks it up. None of the theories that try to explain why humans evolved light skins as they migrated to the dismal climates of the north is altogether satisfactory. Skin cancer is found among people with light skins who expose themselves to ultraviolet by sunbathing. Black people almost never get the disease, but that illness has probably not produced the global trend. First, it is rare even in whites, with about one case per ten thousand people per year. More important, it is a disease of the old. Those who die from it have already passed on their genes, those for colour included.

Without vitamin D, children get rickets, soft and deformed bones. Vitamin D can be made in the skin by the action of ultraviolet light. Under a UV light, whites synthesise a useful dose in half an hour, while blacks take six times as long to do so. Even a few hours in sunshine allows a light-skinned baby to avoid rickets and it is no accident that African babies are lighter than are adults. Perhaps natural selection favoured light skins as man began his long walk from the tropics to the gloom of northern Europe.

But why black skin? Patients treated with UV to combat skin diseases show a sudden drop in vitamin and antibody levels and perhaps black skin protects against this. Another idea is that that black skin allows the peoples of the tropics to warm up at dawn as the sun rises, even if they have to shelter from the heat of the day – when it may act as camouflage in shady places. It is easy to make up stories about how selection may favour certain genes, but none can be taken seriously without experiments.

In fruit flies heat has many genetic effects. It acts on inherited differences in enzyme structure, increases the mutation rate, and even causes 'selfish DNA' to hop around. Humans can regulate their internal temperature quite well so that differences in climate must have a less direct effect. Even so, blood groups and the alternative forms of certain enzymes show north-south trends. Whether these are due to climatic selection is not known.

Humans, like most animals, live on a thermal tightrope. If our body temperature goes up by a few degrees, we die. Molecular biology has illuminated the imminence of thermal disaster. In snails and fruit flies heat shock proteins are switched on when life gets too hot. Sometimes, most of the cell's machinery is devoted to the job. During a fever, our own cells make such proteins. They cluster around delicate enzymes which might be damaged by high temperature. Even a rise of a couple of degrees sets the protective machinery into action. Perhaps people from tropical and temperate climes differ in the sensitivity of the heat shock system. As yet, nobody knows.

Lower animals were once described dismissively described as 'cold-blooded'. They lack the machinery which keeps mammals warm, but many hold their temperature stable by the way they behave. One species of lizard thrives from the deserts of California to the ice caps of the Andes. It keeps its temperature almost the same across this vast range just by moving in and out of the sun. I once invented a paint which fades at a measurable rate when exposed to daylight. To put spots of this onto snail shells shows how long each animal has spent in the sun over a month or so. Snails from hot and cold places behave differently and within a population dark- and light-coloured individuals (which differ in the extent to which they soak up solar energy) also differ in exposure to sunshine. Perhaps the method could be used to study dark- and light-skinned people, too.

Behaviour can be crucial in the control of temperature. Desert lizards cannot stray more than a couple of yards from shade before they die of heat stroke, but are obliged to venture into the sun every few minutes to feed. Some spiders spend half their energies in shuttling between sun and shade. A spider in a place with the right balance of shady and sunny patches can produce far more eggs than one whose home has plenty of food but not enough sunshine. It is easy to forget the importance of behaviour in our own thermal lives. A quick estimate of how much a choice of the right temperature costs the average Briton (or, even more so, the average inhabitant of Chicago) – a bill which includes houses, clothes, central heating, air-conditioning, food and holidays in Marbella or Florida – shows that the spiders are modest in what they spend on keeping comfortable. Warm-blooded we may be, but evolution has forced us into some cold-blooded decisions about how to stay alive in the move from the tropical climates in which our ancestors evolved.

Humans, like most mammals. are adapted to lowlands. They cannot survive for long at over five thousand metres as the amount of oxygen in the air is half that lower down. In the Andes people live at this height. The children of Andean Indian are better able to cope with such conditions than are those of immigrants from the plains. Native highlanders brought up at sea-level are better at extracting oxygen from the mountain air. Perhaps there has been an evolved response to oxygen starvation.

Diet, too, has been an agent of change. the world as a whole only a minority of adults (the population of western Europe included) can digest cows' milk. Most animals (humans before agriculture included) never have the chance to drink milk of any kind after they have been weaned. Its digestion depends on an enzyme that allows the milk sugars to be broken down. If it stays active until adulthood, cows' milk is a useful food. If it does not, milk loses much of its value and an adult who drinks it suffers from wind and indigestion. The relevant gene is rare in much of Africa and in the Far East (which means that the dried milk once sent as food aid to these places was largely wasted). It is much more common in western Europe and in some Africans such as the Fulani of northern Nigeria who herd cattle. Which is the evolutionary chicken and which the egg is not certain. Perhaps the gene was favoured in desert peoples as it allowed them to drink camels' milk to get water. In Europe it may be advantageous because those who have it can extract calcium from cows' milk and avoid rickets. Again, it is easy for imagination to take precedence over experiment.

The best-understood force of selection in humans comes from inherited differences in resistance to disease. Disease is an unavoidable part of existence; and even creatures preserved from the dawn of existence show signs of infection. The games of computer 'life' based on an analogy of natural selection have their sicknesses in the form of computer viruses. Disease has a history and a geography: people have faced different plagues at different times and in different places. Infection is a relentless enemy. It involves creatures who themselves must change in response to the body's defences, or die out, in an evolutionary arms race between ourselves and our diseases. To see what natural selection can and cannot do in response is the task of the next chapter.

Chapter Thirteen

THE DEADLY FEVERS

A fifteenth-century chronicle by the Portuguese explorers of West Africa expresses a bitter complaint: 'It seems that for our sins, or for some inscrutable judgement of God, in all that we navigate along He has placed a striking angel with a flaming sword of deadly fevers.' Three hundred years later, half the Englishmen who went to that part the world died within a year. When Europeans and their African slaves first went to South America, it was the natives' turn to suffer. The population of Mexico dropped from twenty-five million to one million between 1500 and 1600. Some tribes disappeared. The number of Quimbaya in Colombia who paid tribute to the Spaniards was fifteen thousand in 1539, but sixty-nine in 1628. Everywhere, the great killer was infection: malaria, smallpox and typhus. In both New and Old Worlds those who had lived with a disease for many generations survived better than those who experienced it for the first time. There seemed to be inborn differences in resistance between people from different places that at the time seemed almost miraculous. Now, the evolution of defences against disease is the finest example of natural selection in action. The Age of Disease might be (at least for the moment) over, but its genetic consequences will persist for many years to come.

Western society has won a respite in the battle, but throughout recent evolutionary history pestilence has been the greatest killer and the greatest agent of evolution. In the fourteenth century – thirty generations ago – the population

The Deadly Fevers

of England was halved by the Black Death. Death from cold or starvation may be brutal, but at least the enemy is predictable. Bacteria and viruses are themselves alive. They have an ecology, as they need a constant supply of new victims. They can evolve, which leads to a race between natural selection on our survival and that on their ability to infect us. It is an implacable and endless relay race. As soon as one opponent is defeated, another comes along.

Patterns of infection depend on the number of potential victims. As a result, the importance of disease has changed throughout history. The longer an illness has a hold and the more efficiently it is transmitted, the smaller the population needed to allow it to persist. Immunity also plays a part. Some sicknesses must have started long before others. Signs of tuberculosis, which can drag on for decades, can be seen in the bones of those who died tens of thousands of years ago. In contrast, measles is new. It does not last long and is not very infectious. Those who have been infected become immune and cannot be attacked again so that a large population is needed to keep it going. The history of measles is close to those of the many novel ills that have inflicted themselves upon the human species since it began.

In a population which has never been exposed and in which has no immunity, measles can have terrible effects. When it came to Fiji in 1875 (the result of a visit by the King to Sydney) it killed a third of his hundred and fifty thousand subjects. It soon disappeared from the island as it needs a community of at least half a million to persist. Measles may arrive in a place with fewer inhabitants, but cannot maintain itself. In Iceland before the Second World War there were gaps of up to seven years between epidemics. Only after 1945 (when constant movement meant that the Icelanders became part of the European population as a whole) did measles become a continuous problem. Humans have lived in groups of half a million or more for a mere two or three thousand years, so that measles must be a fairly new disease. Its initial impact was much worse than was its effect on populations who had lived with it for many generations.

The constant change in the pattern of infection means that evolution can never rest: far from perfecting us, it is constantly faced with new challenges. Ten thousand years ago, when humans lived in small bands, contagious disease may scarcely have existed. No doubt there were plenty of lice and tapeworms around as their long lives and ability to reinfect their own hosts mean that they do not need many people to keep going. In general the ancient world was a healthy one. People starved, froze or were eaten by tigers instead. Even when an epidemic struck, it was a local matter. The few hunter-gatherers who remain show remnants of this pattern. In the 1950s, tribal groups of Yanomamo differed greatly in the antibodies that they possessed. In some villages, everyone had antibodies to (and must have been infected with) chickenpox. In others, that disease had never arrived but the whole population had once had influenza. The pattern of disease reflected a balance between a chance arrival of a new pathogen and a local epidemic which ended as soon as everyone was immune or dead. The same pattern exists in chimpanzees today. Now, the Yanomamo have joined the rest of humanity and its diseases and have suffered as a result.

With farming, the human population shot up and began to coalesce into one continental mass. A whole new set of disorders appeared. Irrigation helped with the appearance of water-borne parasites such as schistosomes, which are carried by snails. Their eggs have been found in mummies from 1200 BC. Schistosomiasis is still common in Egypt. Many of the biblical plagues were new diseases that took hold as the population of Egypt grew large enough to sustain them.

Some contagions came from animals. The closest relative of measles is the cattle disease rinderpest and measles itself may have evolved from this. A relation of smallpox is found in cows and of sleeping sickness in wild game. Only a small genetic change in the parasites was required to allow them to attack a new host, *Homo sapiens*.

Some maladies came and went and have never been identified. The towns of mediaeval Europe suffered from outbreaks of dancing mania in which thousands took part. Some may have been due to mass hysteria, but the swellings and pain suggest an organic cause. In Italy the affliction was called tarantism and ascribed (wrongly) to the bite of spiders. St Vitus' Dance, with its visions of God, may have been the same illness. The epidemics began in eleventh-century Germany and had disappeared by the seventeenth century. England had its own mysterious and transient illness, the English Sweat, which came and went several times between 1480 and 1550.

The effects were terrifying. It was brought into London by soldiers fleeing the Battle of Bosworth and a month later was at its peak. It killed its victims within a day. The University of Oxford was closed for six weeks. The disease returned several times over the next fifty years and in parts of Europe the mortality was so high that eight corpses were put into each grave. The last epidemic started in Shrewsbury in 1551, killing thousands. Since then the disease has gone. What it was, nobody knows.

Hippocrates, in the fourth century before Christ, was the first to describe symptoms well enough to allow diseases to be diagnosed with any confidence. Ancient Greece had diphtheria, tuberculosis and influenza, but none of his records suggest the presence of smallpox, bubonic plague or measles. Travel soon led to a new set of pestilences. Smallpox was in India a thousand years before Christ, but its short incubation period meant that it killed its carriers

quickly and did not travel well by land. It reached Europe
by sea, with the first epidemic in Rome in 165 BC. Perhaps
the disease helped in the spread of Christianity, as even to
give a sufferer a glass of water can relieve some of the
symptoms. Anglo-Saxon records mention nearly fifty epidemics between 516 and 1087.

Life got even worse in the first large towns. Big cities
are recent things. Before 1800 just one European in fifty
lived in a settlement of more than a hundred thousand
people. There has been movement from the countryside
for a thousand years, but epidemics meant that no city
could sustain its own numbers until the nineteenth century.
London at the time of Pepys had a population of a hundred
thousand, but it needed five thousand immigrants a year
to maintain its population in the face of pestilence.

Plague had killed millions in the centuries before his
day, but its last and worst epidemic was during Pepys'
own lifetime. In December 1664, two Frenchmen died in
Drury Lane. In the following June, Pepys wrote in his
diary: 'This day, much against my will, I did in Drury Lane
see two or three houses marked with a red cross upon the
door, and the "Lord have Mercy upon Us" writ there:
which was a sad sight to me, being the first of the kind
that, to my remembrance, I ever saw. It put me into an ill
conception of myself and my smell, so that I was forced
to buy some roll-tobacco to smell and to chew, which took
away my apprehension.' By the summer of that year, two
thirds of the population of London had fled and the disease
raged throughout England. The cycle of epidemics which
tormented the capital and reached its peak in the Plague
Year of 1665 ended with the replacement of thatched roofs
(and their resident rats) by slates after the Great Fire in
1666. The last European plague was a century later, in the
Balkans. The disease has often been introduced since then,
but has never spread.

Three hundred years ago, England had great cycles of death. Life expectancy fell from forty-two years in the late sixteenth century to thirty years in the seventeenth, returning to the earlier level only in Victorian times. The greatest mortality was in low-lying villages. 'Fevers' were usually blamed. City conscripts impressed into the army did better than the healthier youth from the countryside. The urban soldiery were pinched and weak, but had been exposed to infection so often that they were immune to the diseases which slaughtered their country cousins when forced into crowded barracks.

New contagions continue to appear. As well as AIDS, Africa had another mysterious epidemic in the 1970s, when outbreaks of the deadly (and until then unknown) Ebola fever killed half those infected. Even a trivial change can spark off new illnesses. In the past thirty years, Lyme Disease (named after the village of Lyme, in Connecticut, where it first appeared) has become the most widespread pest-borne disease in the United States, with more than ten thousand cases a year. It causes arthritis and a variety of painful nervous symptoms. The malady is due to a micro-organism which spends part of its time in a tick found on white-tailed deer. A few cases were known a century ago but it did not become common until people moved to the suburbs and were exposed to the deer that flourish there. Nineteenth-century sanitation meant that cities became safer places. But it had a cost. Before sewers, every infant was exposed to a constant small dose of polio virus. Their immune system works well and most became resistant. Once the water had been cleaned up only those few children unlucky enough to come into contact with a sudden dose of the virus got the disease. If the World Health Organisation succeeds (as it hopes to) in eliminating it altogether, any accidental escape might cause a disaster.

For most of the world infection is still a scourge. Ten

million lives a year are lost to measles, and five million to diarrhoea, diseases that could, given the political will, be controlled by vaccines and by clean water. Schistosomes, another parasite that should be simple to contain, attack two hundred million people. Faced with this onslaught by a series of changing enemies, natural selection can never relax. As more is discovered about ourselves, the importance of disease, extant or extinct, looms larger. It may be that much of the mass of human variation is a remnant of past battles against infection and that many of the genetic trends across the globe result from selection by disease today or in earlier times.

Disease itself also evolves. If it did not, its agents would soon be extinct. It once appeared that evolution would inevitably lead to a truce with those infected and that the best strategy for a pathogen would be to keep its host – its homeland – alive. Sometimes, no doubt, this is true. However, the genes that drive a disease alter in their own interest alone. If the most efficient way to increase in number is to kill the patient, then natural selection will provide the means to do so.

Before flush toilets (which, in their early days, fed directly into rivers) cholera was less dangerous. It had to keep its prey healthy for long enough for them to move to another village and to pollute its wells with the bacillus. As soon as one patient could infect hundreds more as his waste poured into a river cholera turned vicious: the victim needed to survive only to reach a water closet. If he died from fluid loss while pumping out millions of bacteria, that mattered not at all.

The evolutionary struggle against malaria is the best illustration of the power of disease in evolution. Three hundred million people are infected and the disease kills two million a year – half of them African children. Almost half the world's population lives in malarious regions and, given the rate of growth in the tropics, the death rate may double within the next thirty years. Increased travel means that the disease can spread at a great rate. More than two thousand cases are imported to Britain each year and, now and again, malaria is transmitted within southern England by a native mosquito. In the United States, with its delta cities, the risk of a return of endemic malaria is real (particularly as mosquitoes have found new places to breed such as in the vast dumps of tyres filled with stagnant water that defile the landscape).

The malady is caused by a single-celled parasite, one of several species of *Plasmodium*, which is transmitted by mosquitoes. Females mosquitoes are deadlier than males, as they drink blood (which they need to make eggs). The parasites are injected from the salivary glands and pass to the recipient's liver. Here, they multiply. One infective cell can produce a thousand descendants. These enter the blood, break into red cells and divide again, digesting the haemoglobin as they do. The *Plasmodium* cells need iron, which they take from the blood. To give under-nourished African children iron supplements can as a result lead to a new eruption of malaria, which had been lying low.

Bouts of fever take place as new waves of cells emerge from the reservoir in the liver. Many of the disease's symptoms are due to the release of iron and other toxic breakdown products as the blood is digested. If the parasites enter the brain there may be a fatal cerebral malaria.

Once the sufferer has been bitten by a mosquito, parasites go into their next phase. Within a man or woman *Plasmodium* has a life of blameless rectitude as it does little but make thousands of identical copies of itself. In the mosquito it has sex. Males and females mature and mate to give new combinations of genes among their offspring. The next generation migrates to the salivary glands, where they are ready to be injected into a human to start the cycle again.

The several species of malaria parasite have an unexpected evolutionary history. Some of their genes are similar
to those found in green plants, so that, perhaps, in the
distant past, their ancestors were related to single-celled
plants (perhaps those whose modern equivalents cause the
'red tides' that kill fish). The DNA of the most virulent
form, *Plasmodium falciparum*, is similar to that of one
which infects birds. Other malaria parasites are closer to
those that attack apes. They may owe their relative mild-
ness to a long history within our relatives.

Falciparum malaria needs a dense population to maintain itself. It probably began ten thousand years ago, when
Africans shifted from the savannah to farming on the edges
of the forests. The symptoms can be recognised in documents from ancient Egypt and China. Hippocrates was the
first to point to its association with wet places. The
swampy area around Rome – the Campagna – was unin-
habited for most of its history because of endemic malaria
and the disease destroyed the prosperity of the coastal cities
of Greater Greece, such as Sybaris and Syracuse. As a result
of malaria the fertile Yangtse basin was abandoned for a
thousand years. The disease spread over the whole world
with the advance of exploration. It was once common
in East Anglia, whose local population were once called
'yellowbellies' after the jaundice caused by chronic infection. It killed King James the First and Oliver Cromwell;
and Sir Walter Raleigh on the scaffold was concerned that
his trembling might be interpreted as fear rather than what
was, ague (or malarial fever).

Hundreds of millions are infected and millions die. Even
so, there seems to be an uneasy coexistence between the
parasite and its host. Evolution has provided dozens of
ways to foil its activities. How humanity copes with the
disease demonstrates the strengths and weaknesses of natural selection. All kinds of defences have appeared. Some

are effective, some less so; and some impose a terrible cost on those who use them.

One of the great puzzles of biology is the mass of inherited variation on the cell surface. It is important, as it prevents people from accepting tissues from each other. But it did not evolve to make kidney transplants difficult. Perhaps such diversity is a relic of a history of natural selection by disease, with particular antigens favoured because they protect against specific infections. For malaria, selection must have been strong as half of the population of West Africa have protective antigens although the most severe form of the illness has been around for a mere five hundred human generations.

Once the *Plasmodium* gets into the red cells there are other defences. Peoples of the Mediterranean and the Middle East bear a mutation which reduces the activity of an enzyme. This makes it hard for the parasite to feed itself, and the cells die, as do their invaders.

Evolution has come up with hundreds more tricks involving the red blood pigment, haemoglobin. In some places in West Africa, up to a third of children carry one or two copies of a mutated haemoglobin known as sickle cell. They have a single alteration in the DNA. This in turn leads to a change in one of the amino acids, the building blocks of the red pigment. When a cell from a carrier of sickle cell is attacked, the haemoglobin forms fibres and the cell collapses, slowing the invader's growth. This is very effective. A child with a copy of the gene has a ninety per cent protection against the disease.

India and the Middle East have mutations of other amino acids which act in much the same way, while Italians, Cypriots and others have evolved more drastic defences. Whole sections of the haemoglobin molecule are deleted. Once again, this slows the growth of the *Plasmodium*. The name of these diseases, the thalassaemias, reflects their distribution, meaning, as it does, the anaemias of the sea (in this case the Mediterranean). The response to malaria can also involve the persistence into adulthood of a haemoglobin normally found in the foetus.

The picture looks pretty confused already. Now that it is possible to use DNA to examine it in more detail, things have got even more complicated. What seemed to be the same defence mechanism in separate places turn out to be genetically quite different. Scores of distinct deletions of bits of the haemoglobin chain are found, as are many different protective cell-surface cues. Altogether, hundreds of mutations have been pressed into service in the struggle. What is more, the same mechanism (sickle cell, for example) has turned up independently in populations a long way apart. Several distinct foci of the sickle gene, each associated with a different set of variants in the adjacent DNA, are known from Africa, with another from India.

A few patches of sickle-cell haemoglobin are found in Europe, where in some places the mutation is carried by people with white skins. One is in the town of Coruche, in central Portugal, where malaria was once common. Most of its inhabitants' DNA resembles that of other Europeans, but the DNA around their sickle-cell gene is of a type only found in West Africa. The Portuguese brought home the first slaves in 1444 and, a century later, the Algarve was filled with Africans and their children, a high proportion of whom had a white parent. Many of these children must have carried sickle cell. This protected them against the local disease so that the African gene flourished and spread although those for black skin were absorbed into the local population and, after hundreds of years, lost from sight.

The malaria story has many lessons for evolution. In any revolution, the mob grabs whatever is at hand to make a rough and ready barrier which, even if it does not stop the forces of repression, slows them down. Natural selection has responded to disease in the same way. Whenever a mutation which might be useful turns up it is used to try and halt the invader. In different places, different genes become available, and the first at hand is used even if it was not the best. The solution which emerges may be wasteful and inefficient, but an ability to 'make do and mend' is characteristic of evolution. It explains why no creature is a beautiful solution to the problems of its own history and why life is, basically, such a mess.

There is a famous anatomical example of the expediency of existence. In all mammals, one of the cranial nerves takes a slight detour around a vertebra in the neck. In giraffes the neck is much extended – but the nerve, far from taking a short cut direct to the brain, goes all the way down to the bottom and back up again. Awkward solutions to an evolutionary dilemma are common and can influence molecules as much as nerves. Perhaps they will explain why most of the structure of DNA also seems to be, to put it bluntly, a shambles.

When faced with an emergency, people often turn to crude solutions that turn out to be expensive in the long run. Evolution does the same. Some of the protective mechanisms against malaria damage those who use them. When the sickle-cell mutation first appeared it was rare, so that almost every copy was partnered by an unchanged gene. This combination protects against infection and leaves its carriers in good general health. As sickle cell became more common, people with two copies of the altered haemoglobin, one from each parent, appeared. They suffer from sickle-cell anaemia, a severe (and sometimes lethal) ailment. Their red cells collapse even when the parasite has not entered, to give a range of symptoms that include brain damage, heart failure and paralysis. In some places, around one child in ten is born with the condition.

That is a high price to pay for protection, but is unavoidable once a population starts using the gene. Some of the other mechanisms (including the thalassaemias) incur the same cost. As more than one person in twenty worldwide carries one or other of these genes, hundreds of thousands of children with inherited anaemias are born each year. This does not add much weight to the idea of natural selection as a benign designer.

Other variants which we now see as inborn disease may themselves be, like sickle cell, relics of a defence against infection (perhaps against illnesses which have now disappeared). Sickle-cell anaemia is found in American blacks, who are not exposed to malaria. If its association with infection elsewhere in the world was not known its presence in that racial group would be a mystery. Other ethnic groups have their own inborn illnesses. One Ashkenazi Jew in thirty is a carrier of the gene for Tay-Sachs Disease. Those who inherit two copies suffer from a fatal degeneration of the nervous system. Families who carry this gene may have had ancestors more resistant to tuberculosis than the average. As TB was common in the European ghettos from whence most of them came, Tay-Sachs might perhaps be the relic of a system of protection against infection. The cost is still being paid by their descendants. Other diseases – such as ankylosing spondylitis, or 'poker spine' – tend to strike people who carry certain cell-surface antigens. Perhaps this too is a relic of natural selection by lost diseases.

Malaria has other attributes which make it a remorseless enemy. Many diseases have been beaten by vaccination. A weakened version of a parasite can persuade the body to produce antibodies which will attack the real thing. The eradication of smallpox is the most spectacular example of this approach. The malaria vaccine has proved a will-of-the-wisp. *Plasmodium* is enormously variable. One of its many surface antigens (which would have to be mimicked by any successful vaccine) exists in fifty different forms. Dozens may be found in just one village. The parasite's sex life makes things worse. Several genes, scattered all over its fourteen chromosomes, produce cell-surface antigens. Every time *Plasmodium* has sex, they are reshuffled into new and unique combinations. Many patients with malaria are infected with more than one strain, so that new mixtures appear all the time. It will be many years – if ever – before malaria goes the same way as smallpox. It is such a subtle and effective opponent that the genes that protect against it will be needed for a long time yet.

Disease may say more about human diversity. Even the ABO blood groups system might result from its actions. The A and B variants differ in just seven bases in the thousand or so that code for them; O has a single DNA base missing part-way down the message, which scrambles all the text from there onwards and removes part of the cell-surface structure coded for by this gene. AB individuals have some protection against childhood diarrhoea and, more important, against cholera, while those with O are more susceptible to that infection (but might be more resistant to malaria). Other genes, too, seem to be associated with resistance. Perhaps ancient illnesses explain a lot of our diversity. Nevertheless, plenty of infections have gone for ever. Optimists claim that the conquest of disease, cold and starvation means that natural selection has come an end. If evolution has one rule, it is to expect the unexpected. New pestilences may appear and cause as much damage as malaria, or those that seem near extinction will stage a resurgence, as has malaria itself.

The history of the battle against disease says useful things about natural selection. Far from designing a simple and effective protection, whenever a straw appears, it is clutched at. Selection acted like a handyman rather than a craftsman. Its products often seem badly, not to say extravagantly, planned and roughly made. If man is indeed made in God's image, malaria does not say much for divine engineering. This haphazard approach has its strengths. Used by engineers or computer programmers it can make subtle and unexpected things. The logic of selection is that of the living world: to produce a complicated design without a designer.

Natural selection has never in its three-billion-year history produced a wheel, let alone a work of art; although it has managed to generate eyes, brains and other organs of great complexity. This is because of its greatest weakness, its plodding approach. A wheel, or a watch, needs some long-term ideas. To make either demands an intellectual leap that is beyond evolution. Natural selection has superb tactics, but no strategy – but tactics, if pursued without thought for cost, can get to places which no strategist would dream of.

Chapter Fourteen

COUSINS UNDER THE SKIN

Nineteen hundred and six was a successful year for the Bronx Zoo. A new exhibit was pulling in the crowds. An African Pygmy – Ota Benga by name – was in the same cage as an orang-utan. The exhibit caused an uproar, not because it was a shameful spectacle, but because it promoted the idea of evolution, that apes and humans were related. After a time, Ota Benga was released, in part as a result of his habit of shooting arrows at those who mocked him. He moved to Virginia, where he committed suicide a few years later.

The Bronx Zoo view of human evolution was once widespread. Linnaeus himself, who first classified animals and plants, put the idea well in 1754: 'All living things, plants, animals and even mankind themselves, form one chain of universal being from the beginning to the end of the world.' Many still see evolution as a smooth progress, a seamless transition from the primaeval slime to New Labour. Linnaeus recognised several distinct varieties among our own species. As well as the yellow, melancholic and flexible *asiaticus* there was *europaeus*, white, ruddy and muscular; *americanus*, red, choleric and erect; and *afer*, black, phlegmatic and indolent.

The groups of humanity were at different stages. Africans were at the bottom, close to the apes, Asians somewhere in between, and white Europeans – needless to say – at the top. Victorian writers did not hesitate to make the idea clear. Robert Chambers, who wrote an influential book on evolution fifteen years before Darwin, claimed that 'Our brain passes through the characters in which it appears in the Negro, Malay, American and Mongolian nations, and finally is Caucasian. The leading characters, in short, of the various races of mankind, are simply representatives of particular stages in the development of the highest or Caucasian type. The Mongolian is an arrested infant, newly born.'

The theory that races are different has a long and ignoble history that has brought misery and death in its wake. It reached into medicine. Most people have seen children with Down's Syndrome, which is due to an error in their chromosomes. This was called by its discoverer, Langdon Down, 'Mongolism' in his 1866 paper 'Observation on an Ethnic Classification of Idiots' for what seemed to him a good scientific reason – these children had slipped a couple of rungs down the evolutionary ladder to resemble a lower form of life, the Mongols. A Japanese friend once told me that in his country the same condition is called Englishism. The idea is ridiculous. Down's Syndrome is due to a mistake in the transmission of a particular chromosome which is found in all groups of humankind and even in chimpanzees.

The history of race illustrates, more than anything else, the limitations of biology. Biologists have been talking – or shouting – about race for years. Ignorance and confidence have gone together. Politicians take scientists less seriously than scientists do, but the story of scientific racism, as it was known, is a grim one.

I have always felt a certain compassion for those whose ability to despise their fellow men is limited by the colour of their victim's skin. Genetics has – and should have – nothing to do with judgements about the value of one's fellow beings. In this sense, the biology of race has no relevance to racism, which is always happy to bend any scientific fact to its perverse ends. The genes do show that there are no separate groups within humanity. This may be reassuring, but should be beside the point. To depend on DNA to define morals is dangerous. Science evolves. It learns more, and theories alter. Our views on human biology have changed and may change again. The same should not be true of attitudes to human rights. Where biology stops and principles begin must not be forgotten.

Humankind can be divided into groups in many ways; by culture, by language and by race – which usually means by skin colour. Each division depends to some extent on prejudice and, because they do not overlap, can lead to confusion. In 1987, a secretary from Virginia sued her employer for discrimination as she was black. She lost the case on the grounds that, as she had red hair, she must be white. She then worked for a black employer and, undaunted by her earlier experience, sued him for picking on her as she was white. She lost again as the court found that she could not be white as she had been to a black school.

Nations, too, differ in how they define their racial affinity. In South Africa just one African ancestor, even in the distant past, once meant ejection from the white race. In Haiti, in contrast, Papa Doc proclaimed his country to be a white one, as almost everyone – dark though their skin might be – had a European ancestor somewhere. Other countries developed fine distinctions based on colour. Latin America once recognised more than twenty races. The off-spring of a Spaniard and an Indian was a mestizo, that of a mestizo and a Spaniard a castizo, a Spaniard and a negro a mulatto, a mulatto and a Spaniard a morisco, a morisco and a Spaniard an albino, an albino and a Spaniard a torna atras and so on in a lengthy, hair-splitting and subjective series.

Races were supposed to be distinct because they descend from different ancestors. Ham, Shem and Japhet, the sons of Noah, were popular candidates. Anthropology began with the search for perfect examples of each lineage, for racial types. Africans, Europeans and Asians were seen as separate versions of humankind. Perhaps, its students thought, every race was once a pure and unpolluted line, secure in its ancestral homeland. Only in modern times was that purity sullied by interbreeding. Race mixture was against nature (exceptions were allowed in emergency, as when Saints Cosima and Damian, with divine help, transplanted a black leg onto a white patient).

If the peoples of today are a confused mix of what was once a series of pure races, it might still be possible to identify perfect specimens of the original groups. That unproven idea led human biology round in circles for centuries in a futile attempt to find divisions into which people could be classified. Its early days were spent in a useless search for homelands and migration routes. Harvard University was at the centre of the search for the archetype. Two suitably discreet nude statues once stood in the Peabody Museum of Anthropology. They were based on measurements made in the 1930s on dozens of male and female students. Average these out, the argument went, and one would produce an image of the ideal Harvard undergraduate – the highest form of human being. A remnant of this philosophy survives in the Miss World Contest whose judges try, and fail, to find an objective definition of the perfect woman.

Racial types were usually identified from skulls. The word 'Caucasian' reflects a claim that the skull which best represented white-skinned people came from the Caucasus Mountains so that – perhaps – the white race had spread from those remote fastnesses. Years were wasted in measuring skulls rather than thinking about what might make them different. The most popular yardstick was the cephalic index, the ratio of the length and breadth of the head. Tens of thousands of crania from different parts of the world were measured in an attempt to sort out their ancestral stocks.

The work was futile. There is no evidence at all that there are, or ever have been, populations whose members all share the same cephalic index. Even worse for the poor craniometers, the skull shape of the children of immigrants to America shifted away from that of their parents towards that of people already there. Its shape is in any case affected by natural selection. Populations from hot places as far apart as Africa and Malaya have similar skull form, which differs from that of Scandinavians or Eskimos. Even if they have different ancestry, they have converged to about the same shape. Natural selection means that shared heads do not prove common homelands.

So obvious seemed the differences between groups that scientists were blinded to their own results. Samuel George Morton in his *Crania Americana* of 1830 measured hundreds of skulls. The differences were, he thought, clear: Caucasians had larger brain cases than Mongolians and Malays, who in their turn were better endowed than Africans and Europeans. When the same specimens were remeasured with modern instruments the differences disappeared. Morton's results were due to the omission of some groups which did not fit his ideas, confusion of males and females, and a failure to correct skull size for differences in body size.

Even so, early workers had enormous confidence in the value of skull shape. Such measurements were used by the Nazis in an attempt to sort out those with Jewish ancestry. The Frenchman Georges Vacher de Lapouge who wrote in 1887 'I am convinced that in the next century millions will cut each others' throats because of one or two degrees more or less of cephalic index' was more correct than he feared.

Races could also be classified by language. The term 'Aryan', which gained such sinister overtones, came from the idea of a talented people, the Arya, who migrated from a homeland somewhere in the east, bringing their inheritance and their language with them. The French writer Joseph Gobineau, the father of modern racist ideology, in his 1854 'Essay on the Inequality of Human Races' wrote that 'Everything great, fruitful and noble in the work of man on this earth springs from the great Aryan family'. He persuaded himself that the Aryans had spread to found the cultures of ancient Egypt, Rome, China and even Peru and that 'all civilisations derive from the white race'.

Thor Heyerdahl's famous voyage across the Pacific in search of the founders of the civilisations of Polynesia can be traced back to Gobineau. They gave rise to a long series of attempts to trace historical links among cultures (such as those of the Celts and the Incas) which share sun-worship, massive stone monuments, and mummies. All were supposed to descend from the Aryans, who were often equated with the ancient Egyptians.

Anthropology is the study of the movement of peoples, genes and cultures. These were once all assumed to be the same thing. To observe one's fellow citizens makes it obvious even to an anthropologist that everyone does not belong to a single racial type: people look different. Difference usually means classification and from there it is a tiny step to judgement. The early evolutionists did not hesitate. Blumenbach, who coined the term 'Caucasian', was glad to show where his sympathies lay. Part of his definition was '... the most beautiful race of men ... Nature has lavished upon the women beauties which are not to be seen elsewhere. I consider it impossible to look at them without loving them . . .' Even Rousseau never suggested that the noble savage was black.

Ninety per cent of the names given to themselves by tribal peoples mean 'men', 'the only men', or 'the best men'; that is, we are human, others less so. The Sioux Indians of North America seem to he an exception. The literal translation of 'Sioux' is snake, or enemy. In fact, this name was given to them by an adjacent tribe (and picked up by the first French settlers). The Sioux themselves call their tribe the 'Lakota' – the humans.

The idea that humanity was divided into distinct lineages of different quality had a disastrous impact. The tie of the philosophy and policies of the Nazis to anthropology, and the desire to return to a lost time of pure races, is clear. The Gesellschaft fur Rassenhygien (Society for Race Hygiene) was founded in 1905. By 1908 all mixed marriages in German South-West Africa (now Namibia) were annulled and those involved in them deprived of their citizenship. Haeckel himself, the champion of *The Origin of Species*, wrote that 'The morphological differences between two generally recognised species – for example between sheep and goats – are much less important than those between a Hottentot and a man of the Teutonic race.' His philosophy ended in disaster.

The ties between biology and the politics of difference that began before Hitler were not broken until many years after his death. Until 1913 the Statue of Liberty really did welcome, as its inscription says, the huddled masses, struggling to be free. In his 1916 book *The Passing of the Great Race* the euphonious American, Madison Grant, echoed many of his fellows when he complained that alien races were being grafted onto the nation's racial stock. With the advice of biologists, President Coolidge was moved to say that 'biological laws tell us that certain divergent peoples will not mix or blend. The Nordics propagate themselves successfully. With other races, the outcome shows deterioration on both sides.'

After determined genetical lobbying, the first Immigration Act was passed in 1924. It set limits to ensure that the ethnic composition of the USA stayed at what it had been in the late nineteenth century. Each country was allowed quota of two per cent of the numbers of its citizens present in the United States in 1890 (when most of that nation's people were from the British Isles, Scandinavia and Germany). The law was very good at keeping Eastern Europeans out and left many to the mercies of the other experiment in race hygiene which soon began there. It was not repealed until 1966. The theory of pure races had cast a long shadow. Its spectre has not yet disappeared. A Hungarian political party campaigned against rights for gypsies in the nineteen-nineties as they were 'a disadvantaged group, to whom the laws of natural selection have not been applied.'

Genetics has at last provided the tools to test the pure race theory. The word 'race' itself is ill-defined. It includes social and political as well as biological criteria. In an attempt to escape a difficulty by renaming it the term 'ethnic group' is sometimes used. Such groups can define themselves. The Scots scarcely existed until they were invented by King George IV, who in 1822 visited Edinburgh and, dressed in a Stuart kilt and a pair of flesh-coloured tights, gave the Scots a national identity they never knew they possessed. It took only the imagination of Sir Walter Scott in devising a native culture to produce a new and potent myth. It was based on the kilt, which, as Macaulay said, 'before the Union, was considered by nine Scotchmen out of ten as the dress of a thief'. The Celts, the larger unit to which the Scots claim allegiance, are themselves an illusion. Celtic culture, defined by artefacts excavated in southern Germany, was hijacked by the French and the Germans as well as the Celtic Fringe as a statement of national worth. In fact, trade had more to do with the spread of Celtic civilization than did sex or conquest.

For ethnic identity what matters is what group we think we belong to. For genes it is not so simple. Perhaps those that count are the ones most visible. After all, people do tend to choose mates of the same skin colour as themselves and this might be important when it comes to the nature of race. The theory of pure races made a definite claim about human groups; that they descended from a series of distinct ancestors. If this is so, and mere appearance represents the remnants of this history, then each race should be distinct in most genes and not just those for skin colour or hair form.

What does the genetic atlas look like? Are shifts in skin colour – the result of a dozen or so genes – matched by parallel trends in the tens of thousands of genes that build a man? The answer is, quite clearly, no.

Everyone can see global trends in colour, hair form and so on. Plenty of less obvious patterns exist but what is behind them is quite unknown. Some patterns are so obvious that they almost beg to be justified in selective terms. In England, the gene for blood group B is rare and is borne by fewer than one in ten people. In central Russia and in west Africa, in contrast, it is common, and up to a third of the population carry that variant. In the rhesus system a marriage between a positive man and a negative woman can be dangerous when the mother's blood react against that of her unborn child, but rhesus negative is common in Europe and Africa (albeit rare elsewhere). It must once have had some advantage that allowed it to spread in the face of this penalty.

Even the imaginative are pressed to explain some other trends in terms of selection. Most westerners have sticky ear wax, but that of most orientals is flaky and dry. And why can most Indians taste the bitter substance PROP while Africans cannot and what causes fingerprint patterns to vary so much across the world?

For none of these is there an explanation: but, as so much of modern medicine depends on genetics we have arrived at the rather unexpected position of knowing more about the patterns of change in humans than in any other animal. Hundreds of functional genes and thousands of variants in the non-coding parts of DNA have been mapped. Most vary in frequency from place to place. The picture that emerges is quite different from that supported by those who believe mankind to be divided into distinct races. Man, it transpires, is the most boring of mammals, varying scarcely at all from place to place. The trends in physical appearance are not accompanied by those in other genes. Instead, the patterns of variation in each system are more or less independent. We would have a different view of race if we diagnosed it from blood groups, with an unlikely alliance between the Armenians and the Nigerians, who could despise the B-free people of Australia and Peru. Gene geography shows that people from different places do not differ much and that colour says little about what lies under the skin.

Imagine that the whole world could be measured for the diversity it contains. The job should be easy enough; after all, its people would all boil down into a soup which would just fill Windermere. The total set can be sorted out among individuals, countries and races to see how it splits up. The analysis, which is based on hundreds of genes in scores of populations, shows that around eight tenths of total diversity, worldwide, comes from the differences between the people of the same country: two Englishmen, say, or two Nigerians. Another five to ten per cent is due to the differences between nations; for example, the people of England and Spain, Nigeria and Kenya. The remainder – the overall genetic differences between 'races' (Africans and Europeans, for example) – is not much greater that between different countries within Europe or within Africa. DNA bears a simple message; that individuals are the repository of most variation. A race, as defined by skin colour, is no more an entity than is a nation, whose personality depends only on a brief shared history.

The notion that humanity is divided up into a series of distinct groups is wrong. The ancient private homeland in the Caucasus – the cradle of the white race – was a myth, as were its equivalents in Egypt or Peru. If, after a global disaster, just one group, the Albanians, the Papuans or the Senegalese, were to survive, most human diversity would be preserved. Humans are uniform creatures, because they evolved so recently. DNA sequence shows that the difference among the races is less than a fiftieth that between man and chimpanzee.

For other animals, race means more. The genetic differences between the snail populations of two adjacent Pyrenean valleys is far greater than that between Australian aboriginals and Europeans. For a snail it makes good biological sense to be a racist, but humans have to accept that they belong to a tediously homogenous species.

The fact that genes can be used to differentiate peoples (as skin colour does Africans and Europeans) is scarcely relevant to how different they are. After all, a forensic scientist can separate two brothers suspected of a crime with a blood sample, although the suspects share half their inheritance. Even a single gene may be a reliable indicator. If a bloodstain at the scene of a crime contains sickle-cell haemoglobin, it is almost certain that the suspect has African ancestry; but if it has the gene for cystic fibrosis (unknown among Africans) then the police should look for a European. Neither observation changes the fact that Africans and Europeans have most of their genes in common.

The issue of differentiability versus difference causes controversy in the legal profession. DNA fingerprints are enormously variable. Confident claims were made about how they would revolutionise forensic science. In one American court, the prosecution described the chance of error as one in seven hundred and thirty eight million million. A single trace of DNA – blood, sperm, or even the saliva spat out onto the shirt of someone in close conversation with a supposed criminal – and the suspect would be identified. There was, it seemed, no room for argument. The case was so persuasive that sometimes judges even refused to hear evidence from the defence.

Now, life looks rather murkier. Of course, even if the test is infallible, the people who make it are not. There have been obvious lapses (such as mistakes in labelling samples). Other technical problems can also lead to difficulties.

The stained bands of the samples to be compared are lined up and compared by eye. The eye is an unreliable instrument, which gives plenty of room for error. Juries, typical as they are of the population as a whole, are bad at understanding risk (which is why the National Lottery does so well) and are much more likely to convict if told that the chance of a random match between defendant and sample is 0.1 per cent than the (identical) figure of one in a thousand. These arguments are the stuff of legal dispute and are no different from the controversies about other forensic tests that hit the headlines. However, forensic genetics faces a deeper problem that arises from evolutionary history.

DNA fingerprints are made up of short sequences of the

message which are repeated again and again. The number of repeats and the position in which they occur varies from person to person. A sample from the scene of the crime is compared with one from the suspect and with others from a panel of innocent donors. Rather like an identity parade, witnesses pick out the criminal from a group known not to have committed the crime.

In the earliest days of DNA fingerprinting the FBI set up a reference group of donors made up of white police officers. To some jurors, if the suspect's fingerprint was more similar to that at the scene than to that of each member of the panel, the case seemed indisputable.

This simple approach faces an evolutionary problem. If an eyewitness had seen – say – a white man committing a crime, and then had to pick out the alleged criminal from an identity parade of blacks, legal eyebrows would be raised. The ethnic group of any suspect must be matched with that of the group with which he is compared.

DNA fingerprints evolve quickly. Those from people of African ancestry are somewhat different from those of Europeans (although the overall racial divergence for this character is not much greater than that for enzymes and blood groups, with nine-tenths of total diversity due to differences among individuals). Imagine a black suspect who is wrongly accused of a crime in fact committed by another black man. His DNA fingerprint is compared to that left at the scene and to those of a panel of whites. Genetic divergence between blacks and whites means that the innocent suspect's DNA may be more like that of the criminal than that of any European; and the innocent man is found guilty.

This has led to controversy in the world of DNA fingerprinting and it is right that it should. In the United States, where legalised murder by the state is common, the issue is one of life and death. The rule in American courts is that scientific evidence may be rejected if it is not generally accepted in the scientific community. Appeal courts threw out convictions for murder and rape because they are not satisfied that DNA fingerprinting is 'generally accepted'. Now, the scientists are ahead, with a survey of individual variation in DNA sequence so extensive that the small racial differences pale by comparison. Even so, the tale of genetics and the law is another reminder that objective knowledge can soon be hijacked by those with a subjective view of how it should be used.

People from different parts of the world may differ but the idea of pure races is a myth. Much of the story of the genetics of race, a field promoted by some of the most eminent scientists of their day, was prejudice dressed up as science; a classic example of the way that biology should not be used to help us understand ourselves. Most of today's biologists feel that the moral issues raised by our own biology – racism, sexual stereotypes, and claims that selfishness, spite and nationalism are driven by genes – are issues of ethics rather than science and that science has nothing to do with how we perceive our fellows. Although it may comfort the liberal conscience to find that genetics reveals few differences among the peoples of the world, this is irrelevant to the issue of racism, which is a moral and political one.

As a result, those determined to dislike one race or another are not much impressed by scientific arguments. I once gave a lecture on race when I was teaching in Botswana. The class was delighted to learn that they were almost the same as the white South Africans who so despised them. At the end of the lecture there was just one question. Surely, a student asked, what you are saying can't be true of the Bushmen; they are obviously different from us.

I admit to a certain despair at that; but it was a useful

reminder that although biology may tell us a lot about where we come from it says nothing about what we are. The dismal history of racial genetics strengthens that belief.

Chapter Fifteen

EVOLUTION APPLIED

Evolution is now a practical subject in its own right although many who use it do not realise what they are doing. Inventors once used an approach close to that of the natural world. For gadgets and life, tinkering works; and can be the means to an unexpected end. Just like the engineers who designed stone tools or steam engines with no understanding of physics, the first farmers developed new crops with no knowledge of heredity at all. Pragmatism led, as always, to progress.

Nowadays, technicians in concrete or metal have a different attitude. They design what is needed with as much scientific theory as is necessary. Applied biology, from agriculture to medicine, has adopted this approach only in the last few years and has begun to advance as much as has transport since Stephenson's *Rocket*. For biology, a new steam age (albeit not yet a space age) is upon us.

A fusion of Mendelism and Darwinism has made agriculture much more productive. The amount of food available per head, worldwide, has gone up in the face of the greatest population explosion in history. In the developing world there is still room for progress as half of all crops are lost to weeds (a figure last seen in Europe in the Middle Ages) and disease can lead to the loss of entire harvests. In Africa, indeed, such is the rate of population growth that – against the world trend – the amount of food produced per person is decreasing. Third-world farming has a long way to go before it catches up. General economic weakness is much to blame, but some of its failure is because it lacks the technology used elsewhere.

Darwin or Mendel would each feel quite at home with most modern agricultural research. In Illinois in 1904 an experiment started in which, each generation, the maize plants most rich in oil were bred from. The work still goes on and, a hundred generations later, the amount of oil has gone up by a hundred times with no sign of any slowing of progress.

Such straightforward applied evolution can do remarkable things, as any cattle-breeder can attest. The 'Green Revolution' took a step further down the genetic road. Its success came from crosses between new and productive stocks of rice and wheat, bred in the Darwinian way, and other lines with stiffer and shorter stalks. Just a few genes were involved. Dwarf varieties were crossed with others with rigid stems. Their descendants were mated with stocks that contained genes for high yield and rapid growth. Plants which combined the best qualities of their parents were chosen and the process continued for several generations. Sex – genetic recombination – did the farmers' job by making new mixtures of genes. It solved a major problem of tropical agriculture, the tendency of rice and wheat to grow tall when fertiliser is used, but to fall over in high winds. One simple trick transformed the rural economies of India and China. In fifty years, planned gene exchange gave a six-fold boost in yield, a figure as great as that at the origin of farming ten thousand years before.

Another refinement of Darwinism involves an increase in the flow of raw material upon which it feeds. To damage DNA can produce new genes ready for use by an alert technologist. Penicillin once depended on tiny amounts of antibiotic made in vast vats of fungus. Breeding from the most productive strains gave a hundredfold increase. The next step did much more: mutations caused by radiation and chemicals led to a new generation of antibiotics, never seen in nature.

An even better way to renew the fuel for selection is to import genes from other species. One of the successes was the new crop triticale, a hybrid between wheat and rye. It can grow in dry places and is of benefit to agriculture in places (such as the American Great Plains) low in rainfall. It demonstrates the gains to be made by even a modest investment in moving genes between species. Another approach is to turn to a domestic plant's untamed relatives, as has been done with wheat itself by crossing with wild grasses that contain genes of value on the farm.

The standard agricultural approach of breeding from the best – evolution writ large – has limits, which are soon reached. Many crops and farm animals can evolve no further because they have used up their genetic reserves and have no source from which to replenish them. The constraint is set by sex: by the fact that to make creatures with new mixtures of genes their parents must mate. In spite of occasional lapses in the plant world, there are strict biological controls as to who mates with whom. The partners must be of different sexes but the same species. A few modest exceptions – triticale being one – are allowed: but to recombine genes, in nature or on the farm, sex is unavoidable. That law much decreased the ambitions of evolutionary engineers because genes that might be useful in improving one form are locked away within another.

Agriculture itself began with some mild infringements of sexual convention. Farmers ameliorated nature by clearing trees to allow vegetation to flourish. Plants that never normally meet came together and, from time to time, hybrids appeared. They contained combinations of genes never seen before. The process goes on. Many mudflats around Britain are covered by a tough grass, a hybrid between a local species and one introduced from America. The new mixture of genes does better in a harsh environment than does either parent, and has become a pest.

Chromosomes show that modern wheat began when two species of grass (each of which is still used for food in the Middle East) hybridised. As on the mudflats, the new cross was more productive than either parent. Soon, another grass crossed with the new recombinant, improving it further. This was the predecessor of every one of the billions of wheat plants grown today. The early farmers had moved chromosomes, genes and DNA from one species to another. They were the first genetic engineers.

Now, science has made sex universal. Molecular biology allows genes to be shifted among lineages which were once quite alien to one another; to make recombinant DNA not by the joint efforts of male and female, but by bypassing the inconvenience of reproduction altogether. Genes can be moved from more or less anywhere to anywhere else. At last, DNA can be used where it is needed, wherever it comes from. The biological rules have been broken and a new era of agriculture is at hand.

Genetic engineering began in bacteria, which have a commendable range of sexual interests. They exchange information in many ways; by taking up naked DNA, by a process of mating rather like that of higher animals and by the use of a range of third parties or viruses. This 'infectious heredity' (which suggests that venereal disease evolved before sex) has been subverted by science. The gene to be engineered (which may be from a bacterium, a plant or a human) is put into a piece of viral DNA with the help of various technical tricks. The manipulated virus plus its fellow-traveller is then used to infect a new host. With luck, the recipient will treat the immigrant DNA as its own and make a copy every time its divides. It can be persuaded to generate vast numbers of duplicates of the engineered gene – and large amounts of whatever it manufactures; pure human proteins, drugs, or other materials.

To cross the sexual divide, deep as it is, between bacteria and the rest of life proved unexpectedly easy. Insulin was once extracted from the pancreas of pigs. The human gene was moved to bacteria and large quantities of the pure protein can now be made. Human growth hormone, too – once extracted with much controversy from the pituitary glands of the dead – is now made in the same way. This avoids a macabre and unexpected problem. A few patients caught a nervous degenerative disease from corpses that carried a virus. Now, the factor VIII gene, too, has been inserted into bacteria and patients are treated with its product.

Genetic engineering can also be used against infectious disease. Jenner could use the cowpox virus to vaccinate against smallpox (an experiment which would fall foul of the most lenient Ethics Committee today) because the viruses share antigens, cues of identity recognised by the immune system as the basis of its response. As a result, antibodies against cowpox protect against smallpox. Cowpox itself can cause problems and even modern vaccines have a small risk of a reaction to the foreign protein. In any case, many diseases (such as leprosy) cannot be helped by vaccination because it is hard to grow their agents in the laboratory.

Some clever engineering gets round the problem. Antigen genes from an agent of disease are inserted into a harmless bacterium, avoiding the risk of infection as the genes for virulence have been left out. Antigens from several sources can be put into the same host to give a single vaccine against many infections. A modified strain of Salmonella (which in its native state can cause food poisoning) is used. The bacterium, with its added antigens, flourishes for a short time in the gut and, by persuading the recipient that he has been infected, ensures that antibodies are made.

Evolution Applied

Some of the tricks are simple. Plants can make copies of themselves from a few cells so that many can be produced from one without sex. It is hard to improve trees by breeding from the best, because it takes so long. Instead, a superior specimen has its tissues broken into single cells. Copies of that super-tree can then be grown to give, within a single generation, a super-forest. In the same way, natural vanilla, once extracted at great expense from a tropical orchid, has been replaced with the same chemical made by cultures of cells grown in the botanical equivalent of a factory farm.

The real promise for farming comes from inserting genes from one species into another. A certain virus causes what is almost a plant cancer: tissues lose their identity and the plant grows up distorted. This crown gall virus is good at picking up foreign genes and has been used to move them into new hosts. The first transformed plant, a strain of tobacco, was made in 1984, to great lack of public interest. A dozen years later, tomato puree made from engineered plants was on sale without much controversy. Then, though, public alarm began; and the 'Frankenstein Food' label was invented, gathering around itself a variety of cranks who claimed, with no evidence, that such foods were harmful to health.

Part of the problem is the word 'engineering', which sounds more of a threat than does the 'domestication' used of the first genetic manipulators. Part comes from the caution of biologists themselves. Thirty years ago they declared a moratorium (soon abandoned) on new experiments until safety rules were worked out. Most important, people are always suspicious of technical fixes; the idea that science can overcome all problems. From nuclear power to Concorde the optimism of engineers has often turned out to be short-lived. For the companies involved, public concern (helped by their own bland assurances about safety and by simple arrogance in refusing to label engineered food) has proved a real problem. Monsanto makes many things (although it has now changed its name to disguise that fact); but became synonymous with a supposed attempt to poison the public. So alarmed is industry that it has set absurd standards of safety. One project used genes from Brazil nuts put into soybeans to provide a certain amino acid. As this is in short supply in the third world it might have saved thousands of children. Instead the project was abandoned as a very few people are allergic to the nut itself. The new plant might have killed one or two Americans a year. The end of the research was greeted as a triumph by the Greens. Other false accusations turn on the supposed dangers of resistance to an antibiotic, kannamycin, used to help pick out which engineered plants have incorporated foreign DNA. Kannamycin is not used in medicine, is widespread in nature, and its use in genetic manipulation is in any case becoming obsolete. Even so, kannamycin has been used as a stick with which to beat those keen to improve food production.

Other complaints, with more weight than all this pseudo-science, are based on fears about the future of the landscape or of farming itself. Many people do not like modern industrial agriculture (in spite of its productivity) and genetically manipulated foods will, without doubt, help it to prevail. It also, say the opponents, makes little sense to manipulate wheat to add to the grain mountain; or to drive peasants from the land to the cities. The Green Revolution itself forced Indian farmers from the land as large companies gained control of seed production.

Much the same happened half a century ago in the American mid-West. In the 1930s new strains of hybrid corn were made by crossing two lineages together. Their sale was controlled by combines who by manipulated the price and put small farmers out of business. Another commercial trick played a part. No longer could a producer use his own seed for the following year because a hybrid plant produces new and unfavourable mixtures among its offspring. Engineered seeds pose the same danger of a harvest of the grapes of economic wrath. Few farmers can bargain with an organisation with a monopoly on the sale of a herbicide-tolerant plant – and the herbicide involved. The companies have threatened to sue those who plant the new seeds in a subsequent year without a new purchase (and have been sued in their turn by clients disappointed by its yield and by others whose own crops are polluted by manipulated pollen). New 'terminator technology' prevents engineered plants from setting seed and – as in the mid-West – forces those who use them to buy new stocks for every harvest.

As is often the case in genetics, much more has been promised by biotechnology than has been achieved (particularly in the third world, where few profits are to be made). Some GM crops have lower yields than others, which has led some farmers to give them up. Such is the storm generated by their use that their potential may be long delayed. Thirty million hectares of land were planted with GM crops in 1998; and a million Chinese farmers used engineered cotton. So alarmed is the public (and so over-priced the seeds) that in the west at least the acreage has been reduced since then.

Most of the brouhaha turns on economics and emotion rather than science. Science, indeed, has got rather lost in the fuss. What might genetic manipulation of plants do, given the chance?

Some of the technology aims to increase the range of places in which particular crops can live, with genes that make them tolerant to salty soil, or high temperature, or shortage of water, or allow growth for a larger portion of the year. The Green Revolution turned on a natural mutation that caused plants to grow less tall than normal. Now the gene involved (which prevents the plant from responding to growth hormones) has been cloned and could be introduced into other crops, to give an instant revolution in unexpected places.

Other genes might fight biological enemies. Many creatures produce natural pesticides as they are at constant risk of attack. Such genes from one species can be shifted into another, to cut down the use of chemical sprays. A pesticide much used by organic farmers is taken from a bacterium, *Bacillus thuringiensis*, which is lethal to many insects. The toxin genes have now been introduced into cotton, reducing the chemicals used on the fields. A related trick inserts a gene that makes the plant resistant to artificial weedkillers. 'Round-Up' is much used by soya-bean farmers. 'Round-Up Ready' plants (which represent about three quarters of all genetically modified crops) have a gene that breaks down the chemical, so that the field can be sprayed to kill the weeds but leave the harvest untouched. Plants can even be 'vaccinated' by introducing a few genes from their viral enemies. When the virus strikes it uses the plant's machinery to make copies of itself. If parts of its own structure are already there, the mechanism is disrupted and the attack fails. Virus resistance has been introduced into rice and peppers, and genes that resist parasitic worms into potatoes and bananas, although none has yet been used on farms.

We grow plants because they make useful things; food, for example. As most plants lack certain amino acids it is hard to stay healthy on a strict vegetarian diet. Much could be done by moving the right genes in and many hopes are pinned on 'golden rice', which has within it a new gene for vitamin A (whose deficiency causes half a million third-world children to go blind each year). Some foodstuffs, such as broccoli, contain anti-cancer substances and the

DNA responsible might be introduced to other species. Plants could even be used as biological factories, with the prospect of using potatoes to make antibodies or other blood proteins. Already, rice can make a human protein used to treat cystic fibrosis and other lung diseases.

Other species might be persuaded to make natural oils for use in plastics or fuel. Another option is to interfere with the DNA of trees to reduce the toughness of the wood and to cut down the amount of energy needed when it is converted into paper. Blue cotton and black carnations are on the horizon. The great hope for agricultural engineers is to introduce genes that allow crops to make their own fertiliser. Clover has evolved an arrangement with certain bacteria. The bugs take nitrogen from the air and turn it into a form which can be used by the plant. In return they gain food and protection. Farmers have long used mixtures of grass and clover that are more productive than either grown alone. To put nitrogen-fixing genes into crops would much reduce the need for fertilizers. The potential rewards are huge. All this may mean that plants may rule and that animals will fade in importance as – perhaps – the salmon-flavoured banana takes over.

To develop such new crops is expensive and the research is, of course, done with profit in mind. It must, like the transistor or the vacuum cleaner, be protected. The first known patent was granted in 1421 in Florence to the architect Filippo Brunelleschi for his invention of a barge with hoisting gear used to transport marble. The idea that inventions need protection spread, and – in spite of attempts to do without in places such as China – is now universal. The law works; and without it capitalism would not have developed.

But what about the idea of patenting life (or, at least, genes)? It seems somehow wrong, but the pass was sold long before the days of DNA technology. In the 1970s it became possible to protect agricultural varieties and in 1980, the US Supreme Court gave the green light to a patent for a bug whose genes had been altered to chew up oil spills. Such creatures were the products of years of work by those who sold them, with a real claim to be inventions, in the legal sense, rather than mere discoveries that cannot be patented.

With life, the boundary between the natural and the invented is soon blurred. Can genes themselves be patented? After all, they evolved and are not products of human ingenuity. In spite of much argument thousands of genes are now under legal protection. The law is still dubious about just how far this should be allowed, and an aggressive attempt to patent segments of DNA without even knowing what they do has failed. Patenting, though, is here to stay. It can, like capitalism itself, be unfair; but, like that economic system, seems unavoidable.

The interesting question is not about ethics, but about who owns the patents. 'Biopiracy' is the theft of genes from the third world. The sums involved are large. Seven of the globe's twenty-five top drugs are derived from natural products; aspirin from willow-bark, a cholesterol-lowering medicine from a Japanese fungus, and cyclosporin, a powerful anti-cancer agent, from a Norwegian equivalent. Those nations have gained from such drugs; but vincristine and vinblastine, developed in the 1960s as a treatment for leukaemia, came from the Madagascan rosy periwinkle. That impoverished land has gained nothing from a trade worth millions (although had it obtained patent cover it might have done so). And what about the anti-cancer chemicals found in Asian corals or the material two thousand times sweeter than sugar made by a West African tree? Those genes will be worth millions when cloned – but who owns them? Some companies are quite blatant in their attempts to cull profit from ancient expertise. Basmati rice is an aromatic (and expensive) variety that has long been used in India and Pakistan. Both governments were outraged to find that, in 1998, the Ricetec Corporation of Texas had filed a patent application for its seeds – and, to add insult to injury, that they had been collected by American scientists invited in to search for new genes that might help feed the third world.

The West itself is dubious about the actions of its citizens. In 1997 the United States Patent Office overturned an attempt to patent the active ingredient of turmeric as an aid to wound-healing as this had long been used as a folk remedy in India. Indeed, the nation's own fingers have been burned. The enzyme used in the polymerase chain reaction comes from a bacterium collected in a hot spring in Yellowstone National Park. The Swiss company that owns the patent makes a hundred million dollars a year in royalties, while the Federal Government (the owner of the spring and presumably of the bacteria) gets not a cent. Now, the Parks Service charges a hundred thousand dollars a go (plus a guaranteed share in profits) for any company that wants to prospect for DNA on its land. Although the claims of wealth from tropical nature may be exaggerated – after all, only one of fifty thousand plants tested by the US National Cancer Institute gave a usable drug – the third world is understandably enraged. Now it is fighting back. Amazonian tribes use the skin of certain frogs as a source of poison for their darts. That substance is a pain-killer if used in minute amounts, and an American company is keen to patent it. But, counter the governments of Ecuador and Venezuela, was not the discovery made by their own people and should not at least some the profits come back to them? The Americans disagree (and are annoyed by the Venezuelans, who have put a stop to the collection of frogs by outsiders).

All this is the stuff of commerce and is as familiar to

students of the history of gold-prospecting as much as to those of gene-mining. Like most of the fuss about manipulated crops, it lies outside science. However, science itself has something to say about the implications of the new technology. Not all of it is reassuring.

The most widespread fear is of the escape of engineered forms and of a new plague unleashed upon the world. Biologists have some standard defences against this concern. Manipulated creatures are likely to be less fit than those which have not been interfered with. After all, if the gene gives its carriers an advantage it might be expected already to have evolved. Most farm animals and plants cannot survive outside farms, which is why the streets are not full of marauding sheep or potatoes. The same is true of bacteria and viruses. Children are injected with live polio virus that has been 'attenuated' to make it less dangerous. Surveys of sewage show that this live virus is constantly escaping. That is the key to its success: even children whose parents do not allow them to be vaccinated are exposed to the viruses excreted by their treated friends. The attenuated virus has never survived in the wild, but depends on a supply of newly treated children. If all engineered organisms are as feeble there is not much to worry about.

Even so, it is worth remembering that every domestic animal is a pest somewhere. Cats wiped out many of New Zealand's birds. Goats have done worse in many places, feral pigs are everywhere in the subtropics and even horses can be a nuisance in California deserts. Plants are even more destructive. Everyone knows about the prickly pear in Australia, but a pretty yellow South African garden plant, the sour-sop, has done even more damage there.

The brash biologist can – and does – argue that we know enough not to repeat such mistakes. Biologists also point out that much of what engineering does is quite natural. Recombinant DNA is made every time sperm meets egg; species are not fixed entities as they evolve from one into another, and – often in bacteria and sometimes in plants – they even exchange genes by natural means. Huge numbers of bacteria are produced, mankind alone excreting ten with twenty-two zeroes after it of the minute creatures each day. Because of mutation many are genetically new and a few, because of the vagaries of reproduction, must include genes incorporated from other species. None has spread and gut bacteria are, in the main, benign.

Viruses give fewer grounds for comfort. Most of the flu epidemics that cross the world each winter begin in China, when the human flu virus picks up genes from those of wild birds. Only when they pass from ducks to pigs to ourselves do the new mixtures cause trouble, but they are salutary reminder of our vulnerability to rare events in distant places.

The release of manipulated organisms was long delayed by such concerns. In California, crops are damaged by frost. As the air cools, patches of ice appear on the leaves around natural colonies of *Pseudomonas* bacteria. One bacterial gene causes this tiresome behaviour. Sometimes it changes by mutation to produce an 'ice-minus' strain that does less harm. Now an artificial ice-minus bacterium is sprayed onto plants and cuts down frost injury by displacing the natives. The gene was moved from a normal bacterium, sections cut out and the altered DNA reintroduced. Although the bacteria are in some senses not engineered at all as the genes come from their own species, the plan caused an uproar. This irritated agricultural researchers. As they pointed out, legal controls would not allow DNA to be moved from a weed to a crop, which is what happened when the first wheat was made. After many battles the release was approved.

During the court cases it emerged that the military had

already played with Californian bacteria. They wanted to know how best to infect people. In the 1950s huge numbers of *Serratia marcescens* bacteria, then assumed to be harmless, were sprayed over San Francisco to see how they spread. Now it is known that *Serratia* can infect those already debilitated by disease and that a number of mysterious infections at the time were due to the bug. Even a natural bacterium which appears to have no ill effects is, it seems, dangerous when placed in unnatural circumstances.

And what if a new gene gets out of its own species and into another? Herbicide resistance genes might get from crop plants to their weedy relatives. For plants like potatoes, with no wild species in the Old World or in North America, that is unlikely, but oil-seed rape and sugar-beet in Britain, and sunflowers in the United States have plenty of local relatives with which they could hybridise. In places where wild turnip and oil-seed rape grow close together as many as one seed in a hundred is a hybrid and many of the plants that emerge are perfectly healthy. The Round-Up resistance gene has been crossed into the hybrids and works perfectly well with no apparent effects on survival. A spray-resistant wild turnip – perhaps the first of many resistant weeds – may be around the corner. Animal genes, too, may stray into unwelcome places. So many fish escape from farms that the genetic structure of North Atlantic salmon has already been damaged by crosses between farmed and local populations. Some plan to move anti-freeze genes from Antarctic fish to their warm-water relatives to farm them in colder and more productive waters. What might happen if escaped tropical fish hybridise with the natives?

To release manipulated beings is to play with the unknown and hence, inevitably, to take a risk. Some scientists suggest that it is so tiny as to be not worth considering. They are still in a phase of technological absolutism. Trust

Evolution Applied

us, they say; but like the engineers who developed nuclear power or drained the Florida Everglades, or the Bourbons, they have forgotten nothing of the successes and learned nothing from the failures of history.

Such enthusiasts disregard the nature of their subject. They claim that the chances of an inadvertent monster are no greater than those of a television made from a random mix of electronic components. In this they echo a familiar creationist argument; that the chances of an organ as complex as an eye arising without divine intervention are the same as those of a whirlwind building an aeroplane as it blows through a factory.

For aeroplanes that is true enough. Those who set safety standards for the first experiments on genetic engineering demanded that the risk be worked out in the same way as in the Boeing factory; if the chance of valve number one failing is one in a thousand, and of valve number two is the same, then the joint chance of both failing at once is one in a million. Such calculations made for the risk of a manipulated virus used to attack caterpillars changing to resemble a relative that attacks humans suggest that the danger to be one in innumerable billions.

Such figures, precise though they can be made to seem, are meaningless, for natural selection is all about assembling almost impossible things; not by instant and improbable leaps but by tiny and feasible steps. Not until the unlikely has been reached, do we notice what evolution can do. Engineered organisms will, like any other being, evolve to deal with their new condition and, in spite of the confidence of their designers, some will cause problems. Low risk is not no risk. It is an economic issue – will the benefits outweigh the costs? For genetically manipulated organisms nobody knows as the experiment has not yet been done. There is, though, a precedent in another much-vaunted piece of biological engineering.

DDT was introduced at the end of the Second World War to control lice. It was a spectacular success. The optimists took charge and used the engineer's approach: with money and technology one can do anything. However, biological bumbling soon triumphed over engineering elegance.

After the conquest of the louse, DDT was sprayed onto malarial mosquitoes. Victory soon seemed imminent. The number of infections fell, in Ceylon from millions to scores. The rot set in as genes for resistance spread. The counter-attack has been so effective that malaria is raging at levels greater than before and the World Health Organisation admits that 'the history of anti-malaria campaigns is a record of exaggerated expectations followed sooner or later by disappointment'. The parasites, too, have subverted attempts to engineer them out of existence and in many places malaria treatments are now useless as the disease organism has evolved means of coping with them. Mutation and natural selection helped both parties survive.

The parasites have a variety of tactics. Chloroquine was developed in the 1940s. Forty years ago it worked almost everywhere. In the 1960s resistance appeared in south-east Asia and South America and has now spread over the tropical world. One defence resembles the mechanism used by cancer to combat drugs. Massive amounts of a transporter protein are made and pump the drugs out of the cell at fifty times the normal rate. Genes that give resistance to other drugs – sometimes several at a time – have also turned up. The Walter Reed Army Institute in the USA screened more than a quarter of a million compounds in the search for a new anti-malarial drug. Only two proved suitable. One was mefloquine, and in Thailand almost all the parasites are now resistant. Medicine is now down to the last remedy, with nothing new in sight. As a result doctors are returning to quinine and to an extract of wormwood (first used in China a thousand years ago), treatments that are toxic and not very effective.

The history of genetic engineering may, when it is written, turn out not to be too different from that of the war against the insects, in which evolution prevailed after initial setbacks. All is not gloom. For some targets, insecticides have worked well and continue to do so. Without them, there would have been no Green Revolution, lice might still be carrying typhus through the poorer parts of Europe and malaria killing even more than it does today. In time, no doubt, economics will prevail over hysteria when it comes to genetically manipulated plants as well. The triumph of ingenuity will not be unalloyed. Only one thing is certain about the new attempts to engineer nature; that nature will respond in unexpected ways. Because living organisms deal with new challenges by evolving to cope, genetic engineers, unlike those who build bridges, must face the prospect that their new toys will fight back.

Chapter Sixteen

THE MODERN PROMETHEUS

Geneticists never use the F-word but often have it turned against them. This chapter takes the subtitle of Mary Shelley's great work. Her monster has been used again and again to berate the efforts of scientists. Frankenstein's creation almost gained a Scottish mate, for his maker journeyed to the Orkneys to manufacture a female for his fierce original. He destroyed it at the last moment: 'Even if they were to leave Europe, and inhabit the deserts of the New World, yet one of the first results of those sympathies for which the daemon thirsted would be children, and a race of devils would be propagated upon the earth who might make the very existence of the species of man a condition precarious and full of terror. Had I the right, for my own benefit, to inflict this curse upon everlasting generations?'

Two centuries later, and two hundred miles south, in the Scottish Borders, was born a lamb which, according to the report in one New World newspaper, at once turned carnivore and devoured her flock-mates. She became the most famous sheep in history. Dolly is not a curse, but has a sweet nature (although, like some biologists – her makers not included – she rushes bleating to the front whenever she sees a camera). Reproduction without sex had hit the headlines. It had in fact been around for some time, but the public was not much interested; even when, in 1985, the genes from a sheep embryo cell were put into a different egg and made a cloned lamb.

Dolly was different. Her birth in 1997 amazed researchers because her genes came from an adult cell that had been persuaded to take a leap back into infancy and to start again. Since then, there have been many more cloned mammals – sheep, mice, cattle and goats – with yet more on the way. The egg that made the sheep Tracy was engineered to make valuable drugs in her milk and there may soon be cows that make human breast milk. Dolly herself has a daughter, Bonnie, made in the traditional sexual way, and the descendants of Tracy and her fellows may grow into factory flocks, worth millions.

And what about the ultimate clone? People have long chosen partners, but now, for the first time, comes the chance of the most perfect choice of all, that of a child in one's image. Quite what that implies is not clear – would a cloned Mozart have written *Don Giovanni*? – and quite why anyone would want to do so is uncertain; but the chorus is against it. The World Health Organisation calls the procedure 'contrary to human integrity and morality', the European Parliament is convinced that cloning '. . . cannot under any circumstances be justified or tolerated by any society, because it is a serious violation of fundamental human rights and is contrary to the principle of equality of human beings as it permits a eugenic and racist selection of the human race, it offends against human dignity and it requires experimentation on humans' and the Vatican is certain that it is 'contrary to moral law' as it is 'in opposition to the dignity both of human procreation and the conjugal union'. Political knees jerk as one when they see the chance for a headline, but Dolly's own progenitor – whose views deserve respect – calls human cloning 'an ugly diversion; superfluous and in general repugnant' (which does not inhibit the seven per cent of Americans who, according to one poll, would be happy to clone themselves).

Cloning, though, is but the latest stage in the manipulation of our reproductive machinery. Most people are ready to accept pregnancy termination on genetic grounds; and, in spite of the concerns about modified plants and animals, have no complaints about gene therapy, should that ever come to fruition. However, to interfere with the next generation, by engineering eggs or sperm or by cloning is, it seems, a step too far.

Even before Dolly, sexual technology had begun to explode. The demand is high. One married couple in six suffers from some failure of fertility, and miscarriages take place in about the same proportion of all pregnancies. At least a million people have been born by artificial insemination, and by 2005 there may be almost as many who trace their origins to a test-tube. Indeed, the chances of reproductive success in such a vessel are higher than those when trying to have a baby by more traditional means. Now, such methods are beginning to change the practice of genetics.

Many inborn illnesses, from PKU to cystic fibrosis, can be treated with some success. Such treatments deal with symptoms, rather than putting right the fundamental flaw – which is no more that what medicine does for most illnesses. Gene therapy gives hope of a cure. In its purest form, it offers the hope of replacing a faulty section of DNA with a normal equivalent, and putting right the problem at source. The idea is a powerful one: and nobody who accepts the necessity of inserting a new heart and lung can quarrel with the idea of replacing a piece of nucleic acid. Whatever its promise, gene therapy has, unfortunately, failed to live up to its headlines.

In principle, the job ought to be if not easy, at least feasible. DNA can be inserted into cells in culture in many ways. Copies made in a laboratory are infiltrated with the help of a virus or wrapped in envelopes of fat that are accepted by the cell as its own. Working genes can even be shot into cells by firing gold spheres coated with DNA from a tiny gun. Twenty years ago there were great hopes that such technology would revolutionise medicine. There have been many claims of success, but only one has much weight. Severe combined immunodeficiency is an inherited failure of the immune system that arises from the absence of a certain enzyme. Children with the condition are kept in a plastic bubble to reduce the chances of infection, and are given bone marrow transplants and injections of the enzyme to help their defences. Cells in which lack the crucial protein have been 'cured' with the appropriate DNA. Several children have been treated with such engineered cells. They are still alive and even go to school. As most of them were also given extracts of the enzyme it is not yet certain that their improved health is due to the gene manipulation.

Whatever this success, all other claims have been premature. Two hundred or so patients with cystic fibrosis have been treated and the best result has been a small and transient improvement in symptoms. The technology has risks of its own. One American patient with liver disease was injected with an engineered virus based on that for the common cold and died as a result. Many others in the trials have not survived (although most were already desperately ill). Some of the most frequent diseases are going to be difficult to treat. To cure sickle-cell would involve targeting tiny numbers of cells deep within the bone marrow, as it is these and not the red blood cells which produce the faulty haemoglobin. For diseases such as muscular dystrophy it might be necessary to deliver a gene direct to millions of separate muscle cells, and to ensure that it is switched on at just the right level of activity.

Molecular biology could be used in medicine in many other ways, some of which have attracted the 'gene therapy' label. Cells can be engineered to carry genes that destroy cancer cells or cause them to stop dividing. It might be possible to introduce DNA that stimulate the immune system's own defences into cancer cells themselves, providing them with the seeds of their own destruction. Another ingenious idea is to insert drug-metabolising genes into such cells, and then to treat them with a chemical that is broken down into a poison – but only in cancer cells. To establish the DNA sequence of a faulty gene also gives the prospect of making 'anti-sense' nucleic acid which binds to the genetic message and blocks it to turn off genes that have gone wrong. All this lies in the future.

The new biology offers more hope for improvements in diagnosis. Molecular probes can detect mutations long before symptoms first appear. Cancer cells often develop unusual antigens on their surface as new genes are switched on. It may become possible to work out the shape of the protein involved and to make a match that sticks to the relevant place. Not only will this show where the damage lies, but if a drug is attached, it may be possible to point a treatment straight at its target.

Engineering might do even more: in theory it could be used to treat generations yet unborn. In mice this has already succeeded. Genes inserted into sperm or egg cells may be passed on. The germ line, as it is known, has been changed. Such 'transgenic mice' are valuable research tools. If genes for a human disease are introduced they can used to study its symptoms (although these may differ those found in humans themselves) and the mice may be used to test drugs. Transgenic mice have been made for sickle-cell anaemia and other inherited illnesses, as have transgenic pigs with some of the genes for human cell surface variation. Their organs – heart and kidney – are about the right size for a transplant and are more acceptable to a human recipient than they otherwise would be. Such pigs look just like pigs; but, to our immune system, resemble a human being. A counterfeit heart has not yet been used for transplantation, but soon may be. As more than a hundred and fifty thousand people die in Britain each year because it is impossible to find a matching organ this may become important in medicine.

Every therapy must work to rules. Everyone has rights to their own body and can decide whether or not to accept treatment. The same logic can be applied to genes. To replace damaged DNA, should that become possible, is not much different from transplanting a kidney and the same choices must be made by the person who receives it. To change genes in sperm or egg is different because it alters the inheritance of someone who has no choice. Many feel that on this and other grounds germ line therapy is unacceptable and have tried to add to the Universal Declaration of Human Rights a statement that everyone has the right to a genetic constitution that has not been changed.

Given that any medical advance is likely to alter the genes of future generations that seems a little too inclusive; and the failure of gene therapy puts the notion for the time being in the field of speculation. However, another set of technologies which once seemed impossible have – in contrast to gene manipulation itself – turned out to be remarkably simple. Their potential use on humans has caused a storm; but that is nothing new.

Genetics outside the uterus allows reproduction to be controlled. It ranges from artificial insemination to surrogate motherhood to germ-line therapy to cloning. Each one was, on its introduction, greeted with horror (and British children born by donor insemination were once defined to be illegitimate because of the objections by the bishops) but most in the end were accepted. However, as Raskolnikov puts it in *Crime and Punishment*: 'Man gets used to everything – the beast!' Philosophers talk of the 'yuk factor', the automatic revulsion about interfering with our reproduction. Forty years ago it was, on just those grounds, illegal in Britain to save eyesight by grafting on a cornea from a dead person. Philosophy, it seems, is not much help to the blind.

The first recorded artificial insemination was by the eighteenth century Scottish anatomist John Hunter who used a syringe to impregnate a woman whose husband had a deformed penis. Since then, to assist the work of nature has become commonplace. Artificial insemination outside the body had to wait until 1978, when sperm met egg in a test-tube to produce Louise Brown. The technology is less simple than it sounds, as eggs in the right stage of development must be retrieved from a potential mother; but, even so, about one in four attempts succeed. After hormone treatment, eggs are sucked from the ovary with a fine needle and fertilised with the relevant sperm. This need not happen at once as eggs can be frozen for later use. After a few divisions, the fertilised egg is returned to the uterus; either of the natural mother or, if she as reproductive problems, into a volunteer. Often, more than one is used (which sometimes leads to several children being born); and, sometimes, the ball of developing cells is screened to check whether it carries a genetic abnormality before deciding to continue. About one British birth in a hundred is a test-tube baby and there are about half a million such children in the world today.

Often, the problem lies with the male. Perhaps his sperm is of such low quality that it cannot penetrate the egg. Sometimes, indeed, it is quite unable to move and cannot escape from the testes. In such cases, sperm can be extracted with a needle and sperm heads injected into the egg. Fertilisation over, the egg is implanted or frozen for later use. Some suggest, indeed, that given the increase in genetic damage in the children of older parents it might be wise for a woman to freeze a sample of her eggs during her teens to ensure the health of future children. For a man the task would be even easier.

All this has led to controversy (including questions as to who might own a dead man's sperm) but is becoming part of medical practice, with hundreds of clinics available across the world. Genetics is often involved, with a check for defects in the fertilised egg. Surrogate motherhood, too, has become common since the first fertilised egg was implanted into an unrelated female in Iceland in 1989. It contains some unpleasant reminders of social reality. In almost every case the surrogate is poorer and less educated than the egg-donor (which, at up to $50 000 a pregnancy in the United States, is not surprising). For all these procedures the yuk factor has been forgotten, but for cloning it remains.

I write as a clone and the son of a clone; and one of the few British citizens entitled to commit incest. A clone, of course: we are all one of those, for the billions of cells that we contain are – each of them – copies of the fertilised egg that made us, reproduced without benefit of sex. My mother, as it happens, is an identical twin (one of two hundred thousand or so in Britain), so that another individual shares all her genes. Her twin – her clone – in turn, has a daughter who is legally my cousin, but in genetic terms a half-sister. The question has never arisen, but there is no legal impediment to marriage, close relative though she is.

My mother and aunt are different individuals, so why the fear of clones? Cloning is, after all, common. It is sex that is rare – to persuade two cells to fuse to make one is, in some ways, the antithesis of reproduction (which, in its clonal version, involves one cell splitting into two). Plenty of creatures, from fungi to lizards, manage without it (and even turkeys can be persuaded to lay eggs without benefit of males). Potatoes are clones and animals can, in principle, be multiplied in the same way. Take an early embryo, split it into pieces and, sometimes, each will grow into an identical twin. This has been successful in rhesus monkeys and, no doubt, could be also be done on ourselves. Sheep and cattle have been split at the eight-cell stage, to give (so far) a maximum of five identical offspring. Cows from the same herd may differ greatly in the amount of milk they yield, and as it takes several years per generation it is much more efficient to clone the champion rather than mating her with some favoured bull. In the 1990s, the method was used by breeders in the United States. It failed because the calves tend, for some reason, to develop much larger than usual and either die or demand an expensive caesarian birth.

Cloning of the Dolly kind is more sophisticated, with the movement of nuclei between cells, but is, after all, just another form of reproductive technology. It follows in the tradition of the great Italian biologist Spallanzani who artificially inseminated a bitch in 1782. Cloning itself – the growth of an organism from an egg containing a foreign complement of genes – began in the 1950s with frogs. They have lazy embryos, because frog eggs are so stuffed with food that they divide to make several thousand cells before they use any of their own genes. Their eggs hence scarcely notice the insertion of another piece of DNA as they are well on in development before genetic information is needed. Sheep wait for just four cell divisions – the 16-cell stage of the embryo – before switching on their genes, while humans DNA becomes active after three division, pigs after two, and mice even before cell divisions begins. This slight delay might explain why sheep proved easier to clone than work on mice (which were recalcitrant about accepting alien genes) had suggested.

Apart from simple vanity or dislike of the opposite sex, cloning might be useful as an aid to the infertile. Perhaps a male is unable to make sperm: and one of his cell nuclei might be inserted into his wife's egg to make a clone. Perhaps one partner has a genetic illness and prefers to use the other's genes to avoid the risk to a child. There are also various eccentric ideas about armies made of cloned copies of some dictator. Whether they would obey orders is another question, and any cloned child, identical as it is to its parent, is likely to prove a particular disappointment should it fail to live up to expectations

None of these possibilities is legal – or feasible – at present. All the fuss about what might be done should be tempered with realism. Cloning, even of sheep, is a complicated business. First an egg must be harvested and the cells that are to provide the nucleus made ready. This is fused to an egg that has lost its own nucleus with – with a Frankenstein touch – a burst of electricity, and stored in the reproductive tract of a second sheep. Those that pass the test are then moved to the surrogate mother herself. Dolly was the only one of three hundred experiments that worked and most cloned cattle, sheep and mice have been born dead or deformed. For humans, for the time being, moving cell nuclei around is just too risky. Even so it seems almost certain that cloning will, at least in one form, become part of medical practic.

In the early 1980s it was found that cells from embryonic mice could be kept alive in the laboratory and that some, instead of moving down the path to adulthood, stayed forever young: ready to develop into any tissue when prompted to do so. They have been kept as perpetual adolescents for up to ten years. The technique involves a certain trickery, with various growth factors added to the culture. These embryonic stem cells, as they are called, when injected into another developing embryo, are happy to develop into blood cells, nerves and so on; or, if they find themselves in the right place, into the precursors of sperm or egg. The recipient grows up as a chimaera; a mixture of cells with different genes – in effect, a mouse with four parents.

Human embryos, too, contain stem cells, but as these are obtained from the extras made after test-tube fertilisation, their use has caused controversy. They could be useful in making skin for burn victims, replacing the damaged nerve cells of those with Parkinson's disease, or even to generate whole organs, either for transplants, or to replace old tissue with new. Most illnesses nowadays are those of old age; and with the promise of such cells in fighting heart disease, cancer and so on as much as half the population might benefit from their use.

Nerve cells from foetuses inserted into the brains of patients with Parkinson's disease can relieve their symptoms of slow movement and rigidity: such juvenile cells can, it seems, change their personalities to adapt to the adult brain in which they find themselves. Even adults have stem cells in parts of the body that, like blood, muscle or liver, often regenerate. Such cells can be retrained to take up new and quite different jobs. Stem cells from the brain or from muscles will, with some urging, make blood cells, while the bone marrow is even more flexible as its stem cells can change into brain, liver and muscle. To inject adult stem cells from the marrow of a healthy patient can strengthen the bones of children with inherited damage to the skeleton and the same approach may reduce the severity of symptoms in people who suffer from Huntington's disease. Perhaps other damaged tissues such as those involved in Alzheimer's disease or diabetes might be helped. In mice, those from a normal embryo injected into a mutant animal lacking part of the sheath of insulation around certain nerves (a structure damaged in multiple sclerosis) make the missing material. Such cells injected into paralysed rats restore movement. Stem cells are rare

– about one in ten billion in the marrow – and not all the news is good; in mice, such cells injected into adults can grow into tumours and it may be necessary to add a suicide gene to kill them off if they turn nasty. Their very malleability may cause problems – who, after all, wants teeth to grow in their brain?

If stem cells pay off, a new era of medicine will begin. Perhaps everyone will keep a store of frozen cells taken at birth in the expectation that they will be needed later to repair an organ that fails with age or disease. On a more modest scale, it should be possible for every hospital to fill a freezer with such things taken from thousands of different people in the hope of having one ready for a match with some future patient who has not stored his own. Even if the hope of new organs is not fulfilled, they might be engineered to make them resistant to anti-cancer drugs so that anyone unfortunate enough to get the disease later in life can be treated with larger doses and his blood-making capacity maintained with material kept in frozen adolescence.

Chimaeras made with the help of stem cells sometimes use them to make not liver or brain but sperm or eggs. Then, all its offspring resemble the stem-cell parent, and, if that individual has been engineered, will carry the inserted gene. That played a crucial part in the tale of Dolly. Remarkable animal as she is, Dolly is, in the end, just a sheep. By adding DNA a small proportion of the millions of cells in a culture dish can be persuaded to take up the alien gene and, with luck, force it to do its job. To insert such transformed stem cells into another animal does to mammals what was once possible only with bacteria. Dolly's successor, Polly, was cloned from cells that contained a human gene for the blood protein missing in one form of haemophilia attached to an on-off switch for sheep milk proteins. Sheep cells can be transformed in this way and, from one engineered beast, a whole herd can grow. Not many may be needed: a thousand animals could satisfy the world demand for the enzyme used to help patients with emphysema, but each may be valued at many thousands of dollars.

The public has an impressive capacity for boredom; and many of the methods used to manipulate genes and produce animals without sex are becoming commonplace. In 1998, Switzerland, where the gothic tale of *Frankenstein* begins, held a referendum on whether to ban gene technology, electrical fusion included, altogether. The motion was defeated and the research has gone on. Perhaps cloning itself will, in a few years, be a standard medical technology.

A passage written in 1818 on the first Swiss genetic engineering experiment: 'With an anxiety that almost amounted to agony, I collected the instruments of life around me, that I might infuse a spark of being into the lifeless thing that lay at my feet. It was already one in the morning, the rain pattered dismally against the panes, and my candle was nearly burnt out, when, by the glimmer of the half-extinguished light, I saw the dull yellow eye of the creature open; it breathed hard, and a convulsive motion agitated its limbs.' That reads better than its modern equivalent: 'The birth of lambs from differentiated fetal and adult cells reinforces previous speculation that by inducing donor cells to become quiescent it will be possible to obtain normal development from a wide variety of differentiated cells'. The report of Dolly's genesis does not have quite the ring of Mary Shelley; but marks the beginning of an era that will tax the most Gothic of imaginations. And what would Dolly's spiritual ancestor, that failed Scottish clone, the Bride of Frankenstein, have thought?

Chapter Seventeen

THE EVOLUTION OF UTOPIA

One reason why science fiction is so boring is that it is nearly all the same. The monsters may differ, but the plots do not. The same is true for most imaginary Utopias. From *The War of the Worlds* to *Planet of the Apes* an alien life form appears, masters the human race, and meets its doom because of its own weakness. Most novels of the future ignore one of the few predictable things about evolution, which is its unpredictability. No dinosaur could have guessed that descendants of the shrew-like beasts that played at its feet would soon replace it, and the chimpanzees who outnumbered humans a hundred thousand years ago would be depressed to see that their relatives are now abundant while their descendants are an endangered species.

Evolution always builds on its weaknesses, rather than making a fresh start. The lack of a grand plan is what makes life so adaptable and humans – the greatest opportunists of all – such a success. That utilitarian approach means that speculations about the future of evolution are risky. As Hegel put it, the greatest lesson of history is that no one ever learns the lesson of history.

In the earliest Utopian novels, from Thomas More onwards, societies of the future were quite different from those of the writer's day. They might have golden chamber-pots; but there imagination ended. The people who urinated into them were much like those who preferred to hoard the metal. After Darwin, Utopia evolved: society stayed the same but people changed instead. Many of the best-known Utopian novels trace their visions of the future to Darwin. Samuel Butler, author of *Erewhon* (called in its first version *Darwin Among the Machines*), shared an education – Shrewsbury School and Cambridge – with the great man and was himself a keen evolutionist (albeit an anti-Darwinian). Aldous Huxley's *Brave New World* owes much of its plot to his biological brother Julian and to their grandfather Thomas Henry Huxley, Darwin's bull-dog. H. G. Wells – whose Utopia, in *The Time Machine*, was based on the evolutionary theme of the human race splitting into two species – himself wrote a biological text-book with Julian Huxley; and, as we have seen, George Bernard Shaw, author of *Back to Methuselah*, was a follower of Galton and appeared on public platforms with him.

Sometimes the link between the utopian novel and eugenics is painfully clear. Shaw felt that 'if we desire a certain type of civilization we must exterminate of people who do not fit into it'. H. G. Wells, in his scientific vision of the world to come, the (now obscure) *Anticipations of the Reaction of Progress upon Human Life and Thought*, published in 1901, wrote in favour of euthanasia for 'the weak and sensual' and of genocide for 'the dingy white and yellow people who do not come into the needs of efficiency'. Many Utopias would not have been comfort-able places for those forced to live in them.

This book has been a tale of how humankind has evolved by the same rules as those that propel less preten-tious beings. Humans are, of course, more than apes writ large. We have two unique attributes: to know the past and to plan the future. Both talents guarantee that our prospects depend on much more than genes. Nevertheless, it should be possible to make some guesses from biological history about what the evolutionary forecast might be.

One pessimistic but accurate prediction is that it means extinction. About one person in twenty who has ever lived is alive today, but only about one in a thousand of the different kinds of animal and plant has survived. Our species is in its adolescence, at about a hundred and fifty thousand years old, compared to several times this for our relatives. Its demise is, one hopes (and in spite of the advances of nuclear physics) a long way away and we can at least reflect about what might happen before then.

The rules of evolution are simple and will not change. They involve the appearance of new genes by mutation, their test by natural selection, and random changes as some, by chance, fail to be passed on. To speculate about the future of each process is to predict human evolution. Will the biological Utopia be like its fictional equivalents; will we continue to evolve as rapidly as we have since our beginnings, or is our evolution at an end?

Humans have interfered with their biological heritage since they appeared on earth. Stone tools, agriculture and private property all had an effect on society and in turn on genes. Many people are concerned that the next phase of history will be one in which genetics makes plans for the future. That asks too much of science. Inadvertent change – evolution by mistake – will be far more important than is any conscious attempt to engineer our own biology.

Even the most determined efforts of doctors, genetic counsellors or gene therapists will have only a small effect on the future. Part of that lies in the healing power of lust: in the desire of people to have children for reasons that have nothing to do with science. More is a matter of arithmetic. For recessive conditions, far more genes are hidden in normal people than in those with disease – a hundred times as many for cystic fibrosis, thousands of times for rarer diseases. Whatever happens to those who receive two copies – death in infancy or by pregnancy termination, or cure by gene therapy – is more or less irrelevant to the future. Social pressure against the genetically unfortunate has decreased. In the 1950s a small minority of achondroplastic dwarves found a spouse, but now more than eighty per cent are married, often to someone else in the same circumstances. They often have children but, even so, the great majority of newborns with the condition appear – as they always have – because of new mutation.

Many inherited diseases appear anew each generation for the same reason. Is, as many dystopians claim, the evolutionary future in danger because of an increase in the mutation rate? H J Muller, who won the Nobel Prize for his discovery that radiation causes mutations, himself wrote a dark novel of the future, *Out of the Night*, in which life has been blighted by the accumulation of genetic damage. Perhaps modern civilisation – with its dubious benefits of nuclear radiation and poisonous chemicals – will damage our genetic heritage. Certainly, such things do alter DNA, but the obvious threats such as man-made radiation and industrial by-products, have a smaller effect than do natural sources such as the radon gas that leaks from granite and the poisonous chemicals found in mouldy food. The Sellafield nuclear power station in the North of England is one of the dirtiest in the western world (and the North Sea its most radioactive body of water). The name of the station has itself mutated from Calder Hall to Windscale to Sellafield in a feeble attempt to calm public suspicion. Compared to other sources of radiation, its effects are minor. Avid consumers of shellfish collected near the discharge pipe (and there are not many of those) receive about as much excess radiation as those who fly from London to Los Angeles and back four times a year and are exposed to cosmic radiation as a result.

A more subtle transformation has had a dramatic effect on the mutation rate. In the western world at least, a change in the age at which people have children means that amounts of DNA damage will drop. The rate of mutation goes up with age and the effect accelerates as the years go on. Most mutations (apart from chromosome mutations, most of which are so damaging that they do not pass to the next generation) take place among males. Fathers of thirty-five do not have a rate much greater than those of eighteen, but after that mid-life moment the incidence of damage shoots up (to a rate twenty times higher in pensioners compared to schoolboys). The more old fathers, the more the genetic damage.

People now live for far longer than in earlier times, allowing mutation to take its toll on a higher proportion of the population. The cancer epidemic in the modern world is confined to older people. Cells that give rise to sperm or egg are also exposed to the destructive effects of age, which is why older parents are more likely to have damaged children. Any change in the age of reproduction will hence have an effect on the mutation rate. If the number of elderly parents goes up, there will be more inherited changes; if it decreases, fewer. Social progress has led to just such a shift. The general picture – which applies to much of the third world as well as to developed countries – is subtle and unexpected.

Before the recent improvements in public health most children died young. Parents started having children when they were themselves youthful and continued until death or the menopause. Throughout history the average number of children per couple has been two, or a little more; as, on the average, it still approximately is. The figure is now reached in a new way: not with perhaps a dozen births accompanied by ten deaths in infancy, but with around two planned and healthy offspring.

A drop in infant mortality means less pressure to have children as an insurance against old age. Contraception allows parents to delay their first child (in Britain now until the late twenties, on average) but then to complete their families quickly. Most people stop soon after they have started. As a result, and although the age at which parents have their first offspring has increased, the number of elderly mothers and fathers has gone down. As recently as the 1920s the average Englishwoman began her final pregnancy at over forty; a figure that has dropped by almost ten years. The occasional births to much older women with the help of technology (and the oldest is in her mid sixties) are so rare as to be insignificant.

Males over thirty-five are the crucial group; but the data for men are harder to gather than are those for their partners. Husbands and wives tend to share, if nothing else, about the same date of birth so that, in spite of a few aged (and anonymous) Lotharios, the figures for women contain most of the information for men. Aristotle advised that girls marry at eighteen, and men at thirty-seven. Although the age difference between the sexes is less than that husbands do tend to be three to five years older than their wives.

The changes in sexual pattern are most obvious in postwar Europe. In Britain, Poland and Switzerland the proportion of mothers over thirty-five – and hence of fathers over about forty, the group most at risk of mutation – dropped from around twenty per cent in 1950 to less than five per cent in 1985. In that year, just one mother in fifty in what was then East Germany was more than thirty-five years old, a figure probably smaller than any time in history. In Ireland the influence of the Church, and the many young men who spent a period working overseas, meant that until a few years ago the only means of birth control was self-denial. Most Irish people did not marry until their late twenties, or later, and until not long ago almost a third of all mothers (and a higher fraction of fathers) were over the crucial age; more than twice the proportion anywhere else in Europe. The number is now much lower (albeit still above the average). There has been some reversal of the trend over the past two decades, with the numbers of mothers over thirty-five increasing from its low point of around one in twenty.

The general picture remains clear: old mothers (and fathers) are rarer than they have been for much of the past. This is bound to have an effect on the mutation rate. Down's syndrome (ten times more frequent among mothers over forty-five than in teenagers) is three times more common in Pakistan (which has almost no family planning) than in Britain, because Pakistani mothers are older than their British equivalents. From the male point of view, in Britain the mutation rate in men is about one and a half times that expected if all fathers were less than thirty, but in Pakistan it is three times this low figure. At the moment, at least, it looks as if the human mutation rate is on the way down. Whether this will continue is not certain, but it puts fears about a new race of mutated monsters into context.

Mutation is the fuel of evolution but, as far as can be seen, evolution rarely runs out of steam. Natural selection, though, is its engine and, like most engines, often speeds up and slows down to face changing circumstances.

Selection is an elusive process and it is more difficult to forecast what its future might be. Nature is always liable to come up – as it has so often before – with a nasty shock. The emergence of the AIDS virus shows the risk of this happening again. Even so, in the western world at least, some of the greatest challenges have gone, because of the control of infectious disease. Once such a disease has disappeared the future of the genes that combat it will change. Cypriots carry the inherited anaemia beta-thalassaemia because it defended their ancestors against malaria. That illness has now disappeared from the island – as, in time, will thalassaemia, with the incidence of carriers dropping by as much as one per cent per generation. In time, and given success in public health, the same will happen to the many other genes that resist the infection elsewhere in the world. Soon, they will remain only as mute and waning witnesses to an ancient past.

Life has also got better for babies. They are important; changes in the survival of adults – essential to individuals as they may be – are of not much relevance to selection because they kill those who have already passed on their genes. What counts for evolution is death before reproduction. What happens to the rest of us is, more or less, beside the point. The history of one inherited character, the weight of newborns, shows just how effective improved conditions at an early age can be in reducing the action of natural selection.

At birth, it pays to be average. Underweight babies, needless to say, survive less well than do others. However, babies heavier than normal are also more likely to die in the first few weeks. In the 1930s about half the babies who died in their first year did so because they were above or below the ideal weight. A difference of just one pound had a large effect. Since some of the variation in this character is genetic, natural selection was at work against genes for extreme birth weight as it had been, no doubt, since our species began.

Now, such selection has almost gone. Improved care means that only very underweight babies, or those much larger than average, are at risk. The intensity of selection has gone down by more than two thirds since the 1950s. Nowadays there is little risk in being a baby of even a kilogram above or below the mean weight. What was once a powerful agent of evolution is on the way out.

Improved child-care has also changed the ratio of the

sexes at the age when people begin to choose a mate. At birth, there are slightly more boys than girls. Boys once had less chance of surviving the hazards of childhood, which led to an almost exact balance of the sexes in the late teens. Now, boys survive almost as well as girls do, so that in future there will be a slight but noticeable excess of young men looking for a mate. If (and many dispute the idea) the differences between men and women in size, or in appearance, are driven by sexual selection, perhaps in years to come that aspect of our evolution will (unlike most of it) advance, to give a generation of taller, hairier and more libidinous males.

There are better ways of looking at the future of selection than just to multiply examples of how it works. Natural selection acts only on differences. If everyone lived to adulthood, found a partner and had the same number of children (whether that number was one, two, or ten) it could not operate. We do not need to know what genes are involved to estimate how important selection might be. Simple changes in the pattern of birth and death reveal its actions in the past, the present and, perhaps, in the future.

In affluent countries, the differences between families in how many people survive have much decreased. This much reduces the power of the evolutionary engine. Ten thousand – even two hundred – years ago, the struggle for existence meant a lot. Skeletons from cave cemeteries show that few lived to be more than twenty. If ancient fertility was like that of modern tribal groups each female had about eight children, most of whom died young. For nine tenths of human evolution, society was like a village school, with lots of infants, plenty of teenagers and a few – probably harassed – adult survivors. Almost every death was potential raw material for selection as it involved someone young enough still to have a hope of passing on their genes. Now, ninety-eight out of every hundred new-born British babies live to the age of eighteen, so that selection acting through the deaths of the young (once its main mode of operation) has almost disappeared.

Not until the past few years have humans lived as long as they are able. For the first time in history, most people die old, perhaps as old as biology allows. Life expectancy has risen from forty-seven to seventy-six years in the past century. Progress has now stopped, for some social classes at least. In the USA in 1979 a white woman of sixty-five could expect to live for another eighteen and a half years. In 1999, the figure was almost the same. In Britain, even if all infectious diseases and all accidental deaths were to be eliminated by government decree, average life expectancy would go up by only a little more than a year. There is still room for progress because of class differences in health. A baby born to an unskilled worker in Britain can expect to live for eight years less than one born to a professional person, a difference which, to our national shame, was until recently increasing. In spite of the effects of class, the prospects for any dramatic improvement in longevity are dim. George Bernard Shaw was wrong. We will not go back to Methuselah.

This is important for the evolutionary future. The increase in the number of old people means that more people die for genetical reasons than in earlier times (in the main because fewer are killed by violence or by infection) but, paradoxically, it also means that selection is weaker. The genes that kill are those for cancer or heart disease, which act late in life. Those who die have already passed on their inheritance. Natural selection is much less powerful on genes such as these than on those that kill the young.

Other changes in the balance of birth and death also reduce its opportunities. Few modern peoples are as fertile as they once were. The Hutterites in North America wish for the largest possible family for religious reasons but even they, living in a healthy society as they do, do not often have more than ten children. For most of history – when families of that size were common – people had as many offspring as possible. Only recently has that number begun to decrease.

The new pattern of existence (with fewer children than ever before but most people lasting until the biological clock runs down) emerged about twenty generations ago, compared to the six thousand or so since we first appeared on earth. As a result, evolution has changed the way it works. Selection nowadays acts more on fertility than on survival.

Differences in fertility among families shot up as birth control became popular. The upper classes adopted the idea well before the lower orders. The French aristocracy caught on and reduced the number of children per marriage from six to two in just a hundred years. The Victorians differed in how fertile they were. Victoria herself did well, and Mr Quiverful, in Trollope's Barchester novels, had a dozen children at a time when other clergymen were discreetly limiting their own families to two or three. Now that birth control is widespread, the difference between families has dropped again, but selection through variation in the number of children born is, for the first time in history greater than that working on the number that survive. As a result, the evolutionary fate of our genes depends more on how many children we choose to have than on the chances of their staying alive.

All the best-understood forces of selection – disease, cold or starvation – act on survival rather than on fertility. The shift in the balance of the two may bring in new and unpredictable evolutionary forces. Perhaps inherited variations in the age of reproduction will become important, as those who mature young squeeze in more generations than those who delay their first birth. There has been a drop in the age at which girls become mature (although, in opposition to this trend, western women now marry five years later than they did half a century ago). What this will do to them is hard to say. A good general rule in biology is that nobody gets a free lunch: success in one walk of life must be paid for by failure in another. Fruit-fly experiments suggest that a shift from high survival towards high fertility involves a trade-off in which those that produce lots of eggs die young. The same may happen to humans.

All this is speculation about details. It is clear than natural selection has ebbed away. Modern India is a microcosm of how evolution has lost its chance to mould the human condition. The continent contains a wide range of cultures, from almost tribal hill-peoples to affluent urbanites. It embodies the history of social change in birth and death over the past several thousand years. Differences among individuals in the various groups in the chances of survival and in their numbers of children show that natural selection has lost eighty per cent of its potential in middle class town-dwellers compared to their fellow-citizens who still follow a tribal way of life.

Weakened – perhaps temporarily – as it may be, there is no reason to think that natural selection will change its tactics. Rather than making a new start by designing ideal solution for a particular problem it will, as always, build on our imperfections. History gives little reason to hope that evolution will act as the agent of human perfectibility. It will never make humanity superhuman.

The raw material of evolution and the power of its prime mover are both running out and, as a result, the rate of change is slowing down. Another shift in modern society is bound to influence our prospects. It has to do with the geography of mating.

For most of history, almost everyone had to marry the girl (or the boy) next door, because they had no choice. Society was based on small bands or isolated villages and marriages were within the group. In many places, populations were stable for many years and, as a result, became quite inbred. Almost nobody moved. The DNA in the brains of American Indians, drowned in a peat bog in Florida, show that people who died a thousand years apart had almost the same genes. There was little migration and the Indians had no option but to marry their relatives.

This pattern persisted in much of the West until a few years ago and still holds in many parts of the world. In most places it is changing. An increase in mating outside the group is the most dramatic shift in the developed world's demographic history. The effect has become stronger and stronger and will have more effect on genetical health than anything medicine is able to do. It will also slow the rate of evolutionary advance.

A few societies once encouraged matings with outsiders. In the Ottoman Empire, talented people were produced by promoting marriages between people from different nations. Their children were seen as 'the fruit of the union of two different species of tree; large and filled with liquid, like a princely pearl'. In South America after the arrival of the Spaniards there was what the invaders described as 'the conquest of the women'. Paraguay – the site of Elisabeth Nietszche's failed genetical experiment – was known as the Paradise of Mohammed, and every Spaniard had twenty or thirty Indian women. The Governor excused this, saying that 'the homage rendered to God in producing mestizos [children of mixed race who were raised as Christians] is greater than the sin committed by the same act'.

Outbreeding is not usually due to deliberate policy. Most of it arises, like so many of the biological events that shape the human condition, as a by-product of social change. Cities and transport play a central part as each provides a larger pool of potential mates than was available in the days of rural solitude.

In the Aeolian Islands off the coast of Italy in the 1920s a quarter of marriages were between first or second cousins. The figure has dropped to one in fifty (and in Italy as a whole is now less than one per cent). Britain, with its lack of a peasant class settled on its own land, has always been more outbred than most of Europe, but increased outbreeding can be seen here, too, with a striking drop in cousin marriages since Victorian times.

Elsewhere, the picture is not so simple. Some societies promote marriages between relatives for economic reasons. They are still frequent in Indian villages, where up to half of all unions may be of cousins, or of uncle with niece. Indeed, among Pakistani immigrants to Britain the incidence of cousin marriage is greater than in their native land, perhaps because of social isolation. Almost half of British Pakistanis of reproductive age are married to a cousin, a proportion higher that among their own parents.

A crude but effective measure of how related one's own ancestors may have been is to ask how far apart they were born. If they come from the same village they may well be relatives, but if they were born hundreds of miles apart this is much less likely. For almost everyone today the distance between the places where they and their own partner were born is greater than that separating their parents' birthplaces. In turn, modern fathers and mothers were almost certainly born further apart than were their own parents. In nineteenth-century Oxfordshire the average distance between birthplaces of marriage partners was less than ten miles. Now, it is more than fifty. In the United States it is several hundred, so that most American couples are almost unrelated. All this shows how much the world's populations are beginning to merge. The most important event in recent human evolution has been the invention of the bicycle.

It will take a long time before the mixing is complete: an estimated five hundred years to even out the genetic differences between England and Scotland – and perhaps even longer to get rid of their cultural contrasts. Although homogeneity is a long away, movement is bound to have an evolutionary effect. No longer will large numbers of children have two copies of a defective piece of DNA because their parents are related. Think of a sexual encounter between an African slave and a white slave-owner in early America. Each has a chance of carrying one copy of certain damaged genes. The most common error in whites is cystic fibrosis, in blacks sickle-cell anaemia. Only children who inherit two copies of either will suffer from inborn disease. Because cystic fibrosis is unknown in Africans and sickle cell in whites the child of a black-white marriage is safe from both.

In many parts of the world immigrant communities are merging with the people already there. Imagine that a tenth of the population of Britain were to immigrate from West Africa (where one person in fifteen is a carrier of the gene for sickle-cell anaemia) and to mate freely with the locals. The number of sickle-cell carriers in the next generation in the new mixed British population would go up by seven times. The incidence of sickle-cell *disease* (which demands two copies of the damaged gene, one from each parent) would drop by ninety per cent compared to the previous situation in the two groups considered together. Most children would be born to parents from the two different peoples, one of whom – the British partner – does not carry the sickle-cell gene. The incidence of the indigenous British problem, cystic fibrosis, would drop by a sixth.

This model of race mixture is simplistic but not unreasonable. In Britain now, about one marriage in thirty is between two people of non-European origin; but a third as many is between a non-European and someone whose ancestors were born in the British Isles. The genes of Black Americans are evidence that there has been mating between Americans of African and of European origin for several hundred years. The effect will be more rapid in Britain, where inter-racial marriage is much more acceptable than it is in the New World. Such shifts may mark the beginning of an age of genetic well-being. Increased outbreeding means that recessive genes will more and more be partnered by a normal copy that masks their effects. Social change will dwarf the efforts of scientists to improve genetic health. In time the mixed populations will reach a new equilibrium and many of the hidden genes will reappear, but this will take thousands of years.

Accidental evolution has shaped the genes of small and isolated populations such as the Boers in South Africa and the inhabitants of Tristan da Cunha. In this new mobile world the chances of a small bottleneck and evolution to happen by accident are small indeed. The third part of the Darwinian machine – random change – has, like the other two, lost most of its power.

The great evolutionary fact of the past three centuries has been the population explosion. By the time of the Pilgrim Fathers, the population of the world was twice that on the first Christmas Day. Since then numbers have increased to reach their present six billion. For all creatures, evolution is slower process than are shifts in distribution or in abundance. Many species go extinct before they have a chance to react to an ecological challenge. For ourselves, some ecological disaster (probably an Einsteinian bang rather than a Malthusian whimper) may mean that speculation about any genetical future is irrelevant.

That future, if it does arrive, will be influenced by local

variations in the rate of population growth. Improvements in health care – and a subsequent increase in population number – always precede a decrease in the number of children that parents choose to have. The delay explains the recent explosion. Claims that the world population will double within a century have proved too alarmist. In most places the shift to the new world – a few, healthy children – has taken place more rapidly than even the most optimistic projected.

In the Pilgrim Fathers' day, European genes gained from population growth. Whites filled the world, while black and Asian numbers stayed more or less the same. Now, the equation has shifted. Growth is at its most rapid in Africa. Most European nations, indeed, are not sustaining their own populations, with the mean number of children per family in Italy and Spain well below replacement level. More than ninety per cent of growth is in the developing world, most of all in Africa. The United Nations estimate that more than ninety per cent of the population rise will be in these regions. Africa shows little sign of a decline in birth-rate. The number of children per woman in East Asia decreased from 6.1 to 2.7 between 1960 and 1990; but in Africa the figures for those years were 6.6 and 6.2 A third of the globe's population may be of African origin by 2050. The AIDS spectre casts some doubt on this figure, but any difference in the relative rate of growth of distinct groups itself means evolutionary change. In the past – during the agricultural revolution, for example – increased numbers led to mass migration. In spite of the political barriers to movement in the modern world, future Utopians may be brown.

Nevertheless, most social changes are conspiring to slow down human evolution. Mutation, selection and random change have all lost much of their power. As a result, the biology of the future will not be very different from that of the past. Economic advance and medical progress may even mean that humans are almost at the end of their evolutionary road, as near to the biological Utopia as they are likely to get. Fortunately, nobody reading this book will be around to see if I am right.

Appendix

A BIBLIOGRAPHIC SKETCH

Trying to keep up with the scientific literature is like running up a down escalator. However much one reads, more and more is published until at last one is forced to give up from mere exhaustion, to be plunged into the Basement of Ignorance. Genetics moves so quickly that it is necessary to sprint upwards to stay in the same place. Although the fundamentals of the subject have not changed, many of the discoveries described here – cloned sheep, the complete human DNA sequence, and much more, were made in the late 1990s and the early months of the millennium.

Papers at the cutting edge soon go out of date. With the exception of a few key research papers, I have not tried to refer to all the sources used in writing *The Language of the Genes*. For those with access to a library, the British journal *Nature* and its American equivalent *Science* publish almost every week new discoveries in human genetics and evolution, accompanied by reviews that put the findings into context. Scientists used to read them from the back (where the job advertisements are). It is a sign of the excitement of genetics that more and more now study *Science* and *Nature* as they are written, from the first to the last page.

Human genetics has many specialist journals. The pace setter is *The American Journal of Human Genetics*. *The Annals of Human Genetics* and *Nature Genetics* are also important. The *American Journal of Physical Anthropology* approaches human evolution from a less genetical angle. *New Scientist* and *Scientific American* provide up-to-date information on genetics and evolution, published almost as it happens. Sometimes, even the newspapers get it right.

Genetics and evolution have inspired a number of outstanding books. One of the best is *The Rise and Fall of the Third Chimpanzee* by Jared Diamond (Vintage Books, London, 1992 and HarperPerennial, New York, 1993). His title comes from the discovery that chimps and humans share most of their DNA sequences. Diamond

The Language of the Genes

uses this to build an engaging history of what we can learn about ourselves from our living relatives. A more sedate account of human genetics is in *The Code of Codes: Scientific and Social Issues in the Human Genome Project,* edited by D. J. Kevles and L. Hood (Harvard University Press, Cambridge, Ma., 1991). Matt Ridley's *Genome: The Autobiography of a Species in Twenty-Three Chapters.* (Fourth Estate, London, 1999) is up to date and comprehensive, taking our chromosomes one by one. John Avise's 1998 book *The Genetic Gods: Evolution and Belief in Human Affairs* (Harvard University Press, Cambridge, Ma.) covers a wider field and puts it into a humanistic context. For a more pessimistic view of the subject's possible dangers – which may perhaps have been downplayed in my own pages – there is *Perilous Knowledge. The Human Genome Project and its Implications* by Tom Wilkie (Faber and Faber, London, 1993). Richard Lewontin lives up to his title (*Biology as Ideology: The Doctrine of DNA*, Penguin Books, London, and HarperPerennial, New York, 1993) with a provocative onslaught directed at the self-righteousness which once pervaded much of human genetics. The confidence of the early geneticists was quite misplaced and Lewontin argues that the same is often true today.

Those not put off by scientific terminology – and genetics is plagued with jargon – should try Mange, E.J. and Mange, A.P (1998) *Basic Human Genetics* (Sinauer, Sunderland, Ma.), Strachan, T. and Read, A.P. (1999) *Human Molecular Genetics* (Wiley) or, for a comprehensive treatment, Vogel, F. and Motulsky, A. G. (1997): *Human Genetics – Problems and Approaches* (Springer-Verlag, NY). A wealth of internet links leads to aspects of modern genetics. The main data-base for inherited disease is OMIM (On-Line Mendelian Inheritance in Man) at *http://www3.ncbi.nlm.nih.gov.8o/Omim/.* It contains an encyclopaedic, up-to-date and highly technical account of human inheritance. The Human Genome Project has its own web-page at *http://www.nhgri.nih.gov./HGP/* which is matched by a British equivalent from the Sanger Centre in Cambridge: *http://webace.sanger.ac.uk.*

My own jointly-edited book *The Cambridge Encyclopedia of Human Evolution* (eds. Steve Jones, Robert Martin and David Pilbeam, Cambridge University Press, 1992) has articles on human and primate palaeontology, comparative anatomy, anthropology and genetics. Many of these subjects are covered in more depth by *How Humans Evolved* by R. Boyd and J.B Silk (W. W. Norton and Company, 1997) and *Principles of Human Evolution: A Core*

Appendix

Textbook by R. Lewin (Blackwell Science, Oxford, 1998). For engaging tales about evolution and its eccentrics one can do no better than *The Encyclopedia of Evolution* by Richard Milner (Facts on File, 1990). I have used language as a metaphor for evolutionary change and *The Cambridge Encyclopedia of Language* by David Crystal (Cambridge University Press, 2nd Edition, 1997) is a witty and complete introduction to linguistics for readers who, like me, come from outside the subject.

For those who wish to pursue further the points raised in individual chapters, the following list may be of some help. It is far from exhaustive and, because of space constraints, many topics covered in the text are not referred to here.

INTRODUCTION: THE FINGERPRINTS OF HISTORY

Keynes, M. 1993. *Sir Francis Galton, FRS. The Legacy of His Ideas*. Macmillan, London.

Kevles, D. 1986. *In the Name of Eugenics. Genetics and the Uses of Human Heredity*. Penguin, London, and University of California Press, Berkeley.

Proctor, R. N. 1988. *Racial Hygiene: Medicine under the Nazis*. Harvard University Press, Cambridge, Ma.

CHAPTER ONE: A MESSAGE FROM OUR ANCESTORS

Burley, J. 1999 (ed). *The Genetic Revolution and Human Rights*. Oxford University Press, Oxford.

Bowman, J.E, and Murray, R.F. 1990. *Genetic Variation and Disorders in Peoples of African Origin*. Johns Hopkins University Press, Baltimore.

Esteban J. et al. 1998. Estimating African-American admixture proportions by use of population-specific alleles *American. Journal of Human Genetics*. 63:1839–1851.

Oner, C. *et al*. 1992. Beta S haplotypes in various world populations. *Human Genetics* 89: 99–104.

Neel, J. V. et al. 1988. Protein variation in Hiroshima and Nagasaki: tales of two cities. *American Journal of Human Genetics* 43: 870–893.

CHAPTER TWO: THE RULES OF THE GAME

There are many introductory texts in genetics which deal with the rules of inheritance. One of the most comprehensive and up-to-date is *An Introduction to Genetic Analysis* by A. J. F. Griffith et al.

(7th Edition, 2000; W. H. Freeman). *Genetics: Principles and Analysis* (Jones and Bartlett, 4th Edition, 1998) by Daniel L. Hartl and Elizabeth W. Jones is impressively clear, while my own *Introducing Genetics* (with Borin van Loon: Icon Books London, 2000), if nothing else, lives up to its title.

CHAPTER THREE: HERODOTUS REVISED

The cystic fibrosis story is described in Tsui, L.-C. and Buchwald, M., 1991. Biochemical and molecular genetics of cystic fibrosis. *Advances in Human Genetics* 20: 153–266. As gene mapping is proceeding so quickly it is difficult to keep up to date. A sampler of recent work includes:

Chinnery, P. F. and Turnbull, D. M. 1999. Mitochondrial DNA and disease. *Lancet* 354 supp 1: 17–21; Dunham, I., et al. 1999. The DNA sequence of human chromosome 22. *Nature* 402: 489–495; Little, P. 1999. The book of genes. *Nature* 402: 467–468; and Hattori, M. et al. 2000. The DNA sequence of human chromosome 21. *Nature* 405: 311–219. The web-sites for the Human Genome Project (see above) are filled with the latest information on human gene mapping.

Weiss, K.M. 1998. Coming to terms with human variation. *Ann. Rev. Anthropol.* 27: 273–300

Wright, A.F., Carothers, A.D. and Pirastu, M. 1999. Population choice in mapping genes for complex diseases. *Nature Genetics* 23: 397–404

CHAPTER FOUR: CHANGE OR DECAY

Ames, B.N., Profet, M. and Gold, L.S. 1990. Dietary pesticides (99.99% all natural). *Proceedings of the National Academy of Sciences* 87: 7777–7781.

Crow, J.F. 1997. The high spontaneous mutation rate: is it a health risk? *Proceedings of the National Academy of Sciences* 94: 8380–8386.

Ferguson, L. R. 1999. Natural and man-made mutagens and carcinogens in the human diet. *Mutation Research* 443: 1–10.

Garner, C. 1992. Molecular potential. *Nature* 360: 107–2.08.

Green, P. M., et al. 1999. Mutation rates in humans. I. Overall and sex-specific rates obtained from a population study of haemophilia B. *American Journal of Human Genetics* 65: 1572–1579.

Appendix

CHAPTER FIVE: CALIBAN'S REVENGE

Gayán, J et al. 1999. Quantitative-trait locus for specific language and reading deficits on chromosome 6p. *American Journal of Human Genetics* 64:157–164.

Stinson, S. 1992. Nutritional adaptation. *Annual Review of Anthropology* 21: 143–170. (Diet, genetics and health).

Bouchard, T. J. et al. 1990. Sources of human psychological differences: the Minnesota study of twins reared apart. *Science* 250.: 223–228.

Holgate, S. et al. 1999. The epidemic of allergy and asthma. *Nature* 402: supp B2-B39.

Owen, M. J. and Cardno, A.G. 1999. Psychiatric genetics: progress, problems and potential. *Lancet* 354 suppl 1: 11–14.

O'Donovan, M. C. and Owen, M. J. 1999. Candidate-gene association studies of schizophrenia. *American Journal of Human Genetics* 65: 587–592.

CHAPTER SIX: BEHIND THE SCREEN

Holtzman, N. and Shapiro, D. 1998. Genetic testing and public policy. *British Medical Journal* 316: 852–856.

Kinmonth, A. L., et al. 1998. The new genetics: implications for clinical services in Britain and the United States. *British Medical Journal* 316: 767–770.

Krynetsti, E .Y. and Evans, W. E. 1998. Pharmacogenetics of cancer therapy: getting personal. *American Journal of Human Genetics* 63: 11–16.

Lenaghan, J. 1998. *Brave New NHS? The Impact of the New Genetics on the Health Service* Institute for Public Policy Research, London.

Mao, X., 1998. Chinese geneticists' views of ethical issues in genetic testing and screening: evidence for eugenics in China. *American Journal of Human Genetics*. 63: 688–695.

Marteau, T. and Croyle, R. T. 1998. Psychological responses to genetic testing. *British Medical. Journal* 316, 693–393.

Rothenberg, K.et al. 1997. Genetic information and the workplace: legislative approaches and policy challenges. *Science* 275: 1755–1756.

CHAPTER SEVEN: THE BATTLE OF THE SEXES

Maynard Smith J. 1988. *Evolutionary Genetics*. Oxford University Press. (The origin and maintenance of sex.)

Haig, D. 1997. Parental antagonism, relatedness asymmetries, and genomic imprinting. *Proceedings of the Royal Society of London, Series B*. 264: 1657–1662.

Hurst, L. 1991. Intragenomic conflict as an evolutionary force. *Proceeding of the Royal Society of London Series* 248: 135–148.

Fisher, H. E. 1992. *Anatomy of Love: The Natural History of Monogamy, Adultery and Divorce*. W. W. Norton, New York, and Simon & Schuster, London. (Comparative behaviour of primates).

Johnson, A. M. et al. 1992. Sexual lifestyles and HIV risk *Nature* 360: 410–413. (Male and female mating frequencies.)

Mulder, M. B. 1991. In *Behavioural Ecology*, ed. J. R. Krebs and N. B. Davies (Blackwell, Oxford), pp.69–98. Human behavioural ecology. (Tribal peoples and male success in relation to wealth.)

CHAPTER EIGHT: CLOCKS, FOSSILS AND APES

Gebo, D.L. et al 2000. The oldest known anthropoid postcranial fossils and the early evolution of higher primates. *Nature* 404: 276–278.

Goodman, M. 1999. The genomic record of humankind's evolutionary roots. *American Journal of Human Genetics* 64: 31–39

Kay, R. F., Ross, C and Williams, B. A. 1997. Anthropoid origins. *Science* 275: 797–804.

Richmond, B.G and Strait, D.S. 2000. Evidence that humans evolved from a knuckle-walking ancestor. *Nature* 404: 382–385.

Ruvolo, M. 1997. Molecular phylogeny of the hominoids: inferences from multiple independent DNA sequence data sets. *Molecular Biology and Evolution*. 14: 248–265.

Klein, R. G. 1999. *The Human Career: Human Biological and Cultural Origins* (2nd Edition). University of Chicago Press.

Walter, R.E. *et al*. 2000. Early human occupation of the Red Sea coast of Eritrea during the last interglacial. *Nature* 405: 65–69.

Trinkaus, E. and Shipman, P. 1993. *The Neandertals: Changing the Image of Mankind*. Jonathan Cape, London.

Appendix

CHAPTER NINE: TIME AND CHANCE

Lasker, G. W. 1989. Surnames and Genetic Structure. Cambridge University Press, Cambridge

Ober, C, Hyslop, T. and Hauck, W. 1999. Inbreeding effects on fertility in humans: evidence for reproductive compensation. *American Journal of Human Genetics* 64:225–231

Chagnon, N. A. 1972. In *The Genetic Structure of Human Populations* ed. G. A. Harrison and A. J. Boyce. Clarendon Press, Oxford. Tribal Social organisation and genetic differentiation.

Bittles, A. H., et al. 1991. Reproductive behaviour and health in consanguineous marriages. *Science* 252: 789–794.

Williams, E.M. 1986. In *Genetic and Population Studies in Wales* (ed. P. S. Harper and E. Sunderland, University of Wales Press, Cardiff) pp. 186–211. Genetic studies of Welsh gypsies.

Stine, O. C. and Smith, K. D. 1990. The estimation of selection coefficients in Afrikaners: Huntington's disease, Porphyria variegata and lipoid proteinosis. *American Journal of Human Genetics* 46: 452-458.

Jones, J. S. and Rouhani S. 1986. Human evolution: how small was the bottleneck? *Nature* 319: 449–450.

CHAPTER TEN: THE ECONOMICS OF EDEN

Chikhi, L. et al. 1998. Clines of nuclear DNA markers suggest a largely Neolithic ancestry of the European gene pool *Proceedings of the National Academy of Sciences* 95: 9053–9058.

Corte-Real, H.B. et al 1996. Genetic variation in the Iberian Peninsula determined from mitochondrial sequence analysis. *Annals of Human Genetics* 60: 331–350.

Hoffecker, J. F. et al, 1993. The colonization of Beringia and the peopling of the new world. *Science* 259: 46–53.

Lev-Yadun, S, Gopher, A. and Abbo, S., 2000. The cradle of agriculture. *Science* 288: 1602–1603.

Macaulay V. et al. 1999. The emerging tree of West Eurasian mtDNAs: a synthesis of control-region sequences and RFLPs. *American Journal of Human Genetics* 64:232–249.

McCorriston, J. and Hole, F. 1991. The ecology of seasonal stress and the origin of agriculture in the Near East. *American Anthropologist* 93: 46–69.

Cohen, M. N. and Armelagos G. J. 1991. *Paleopathy and the Origins of Agriculture*. Academic Press, London.

Simoni, L. et al. 2000. Geographic patterns of mtDNA diversity in Europe. *American Journal of Human Genetics* 66: 262–278.

CHAPTER ELEVEN: THE KINGDOMS OF CAIN
Park, T. 1992. Early trends towards class stratification: chaos, common property and flood recession agriculture. *Journal of the American Anthropological Association* 94: 90–117.
Griffeth, R. and Thomas, C. G. 1981. *The City-State in Five Cultures*. ABC Clio, Santa Barbara and London.
Piazza, A, et al. 1988. A genetic history of Italy. *Annals of Human Genetics* 52: 203–213.
Comas, D. et al. 1998. Trading genes along the Silk Road. *American Journal of Human Genetics* 63 1824 1838.
Cavalli-Sforza, L. L. 2000. *Genes, People and Language*. Farrar, Straus & Giroux, NY.

CHAPTER TWELVE: DARWIN'S STRATEGIST
Fleischer, R. C. and Johnston, R. F. 1984. The relationships between winter climate and selection on body size in house sparrows. *Canadian Journal of Zoology* 61: 405–410.
Foley, R. 1987. *Another Unique Species: Patterns in Human Evolutionary Ecology*. Longman, Harlow, Essex. (Climate, diet and human differentiation.)
Guglielmino-Matessi, G. G. et al. 1979. Climate and the evolution of skull metrics in man. *American Journal of Physical Anthropology* 50: 494–564.
Harvey, C B et al 1998. Lactase haplotype frequencies in Caucasians: association with the lactase persistence/non-persistence polymorphism. *Annals of Human Genetics* 62: 215–223.
Baker, H. T. 1992. In *The Cambridge Encyclopedia of Human Evolution*, eds. Steve Jones, Robert Martin and David Pilbeam (Cambridge University Press). Human adaptations to the physical environment, pp. 46–51.

CHAPTER THIRTEEN: THE DEADLY FEVERS
Crosby, A. W. 1993. *Ecological Imperialism: The Biological Expansion of Europe*, 900–1900. Cambridge University Press/Canto Books.
Kiple, K. F. (ed.) 1993. *The Cambridge World History of Disease*. Cambridge University Press, Cambridge.
Kwiatkowski, D. 2000. Genetic susceptibility to malaria getting

complex. *Current Opinions in Genetics and Development* 10: 320–324.

Waters, A. P. et al. 1991. *Plasmodium falciparum* appears to have arisen as a result of lateral transfer between avian and human hosts. *Proceedings of the National Academy of Sciences* 88: 3140–3144.

Desowitz, R. S. 1993. *The Malaria Capers: More Tales about Parasites and People*. W. W. Norton and Company, New York and London.

Stearns, S. (ed.) 1998. *Evolution in Health and Disease*. Oxford Univ. Press.

CHAPTER FOURTEEN: COUSINS UNDER THE SKIN

Baker, J. R. 1974. *Race*. Oxford University Press.

Banton, M. 1987. *Racial Theories*. Cambridge University Press.

Esteban, J. and others 1998. Estimating African American admixture proportions by use of population-specific alleles. *American Journal of Human Genetics* 63: 1839–1851.

Malik, K. 1996. *The meaning of Race: Race, History and Culture in Western Society. Macmillan, London.*

Tapper, M. 1999. *In the Blood: Sickle Cell Anemia and the Politics of Race*. University of Pennsylvania Press, Philadelphia.

CHAPTER FIFTEEN: EVOLUTION APPLIED

Bud, R. 1991. *The Uses of Life: A History of Biotechnology*. Cambridge University Press.

Fowler, C. and Mooney, P. 1990. *The Threatened Gene: Food, Politics and the Loss of Genetic Diversity*. The Lutterworth Press, Cambridge.

Cooper, C et al 1999. A completely GM issue. *The Biochemist*, Oct 99 9–39.

Hails, R.S. 2000 Genetically modified plants – the debate continues. *Trends in Research in Ecology and Evolution* 15: 14–18.

Kishore, G.M and Shewmaker, C. 1999. Biotechnology: Enhancing human nutrition in developing and developed worlds. *Proceedings of the National Academy of Sciences* 96: 5968–5972.

CHAPTER SIXTEEN: THE MODERN PROMETHEUS

Beaudet, A. L. 1999. Making genomic medicine a reality. *American Journal of Human Genetics* 64: 1–13.

Bobrow, M and others. 1999. Molecular medicine. *Lancet* 354

Suppl. 1: 1–37.Gosden, R. 1999. *Designer Babies: the Brave New World of Reproductive Technology.* Gollancz, London.

Wilmut, I., Campbell, K. and Tudge, C. 2000. *The Second Creation: the Age of Biological Control by the Scientists who cloned Dolly.* Headline, London.

Anderson, W. F. 1997. Human gene therapy. *Science* 256:808–813.

Bell, J. 1998 The new genetics in clinical practice. *British Medical Journal* 316: 618–620.

Collins, F.S. et al. 1999. The chipping forecast. *Nature Genetics* 21 supplement.

CHAPTER SEVENTEEN: THE EVOLUTION OF UTOPIA

Vogel, F. 1992. Risk calculations for hereditary effects of ionizing radiation in humans. *Human Genetics* 89: 117–146.

Modell, B. and Kuliev A. M. 1989. Impact of public health on human genetics. *Clinical Genetics* 36: 186–298.

Ulizzi, L. and Terrenato, L. 1992. Natural selection associated with birth weight. VI. Towards the end of the stabilizing component. *Annals of Human Genetics* 56: 113–118.

Reddy, B. M. and Chopra, V. P. 1990. Opportunity for natural selection among the Indian populations. *American Journal of Physical Anthropology* 83: 281–296.

Olshansky, S. J. et al. 1990. In search of Methuselah: estimating the upper limits to human longevity. *Science* 150: 634–640.

Bittles, A. H. et al. 1991. Reproductive behaviour and health in consanguineous marriages. *Science* 151: 789–794.n.

INDEX

Index

Index

Index

Index

cancer 233
colour 232–4, 254–5, 261
slave trade 27–31, 248
sleeping sickness 241
smallpox 238, 241, 251, 272
smell, source of identity 23, 183
Smith, Adam 191
Smith, Sydney 141
smoking, dangers of 102–4, 120
snails ix 186, 234–5, 263
Social Darwinism 108
Society for Race Hygiene 259
South Africa
  Afrikaner people 186–9
  Boers 314
  Cape Coloured people 266
  Klasies River site 162
  racial identity 255, 259
South America
  early human settlement 197–9, 202
  Huntington's Disease 188–9
  resistance to anti-malarial drugs 284
  slave trade 27–31, 238
  Spanish conquests 153, 238, 311
  tribal DNA fingerprints
  see also American Indians;
    Argentina; Brazil; Paraguay
Spain
  South American conquests 210
  surnames 178
  twin births 105
sparrows 224–5, 231–2
speech *see* language
Spencer, Herbert 6, 96
sperm *see* eggs and sperm
spiders 235
spina bifida 99
Statutes of Kilkenny 214
sterilisation 9, 126, 128

Sumerian civilisation 159–60, 211
sunlight
  human desire for and cancer risk 229–30, 233
  temperature control 232–3
Sweden 121
Switzerland
  family names 175–6, 179–80
  lowering of marriage age 304
Syria, early human settlement 200–1

Taiwanese, genetic clues of 197
Tasmania, isolation 195–6
taste, inherited abilities 22–3
Tay-Sachs Disease 250
technology and tools, early
  development 192–5, 198, 301
thalassaemia 99, 126, 128, 247–8, 250, 305–6
Thatcher, Margaret 17
tobacco
  to make antibodies 103–4
  see also smoking
tongue, ability to roll into tube 21
tools *see* technology and tools
Totem and Taboo (Freud) 185
transplantation 247, 290–1
Tristan da Cunha 187–8, 314
triticale crop 270
tuberculosis 239, 241
Turkey
  exchanges with Greeks 213
  language and genes 216, 218
turmeric 279
Tutankhamun, Pharaoh 31
twins, identical and fraternal 105–8, 135, 293
typhus 238

Danah Zohar

The Quantum Self

'Physics without fuss.' Jeanette Winterson, *Sunday Times*

This controversial examination of the relationshp between quantum mechanics and the nature of consciousness turns received scientific wisdom on its head. Zohar shows how old-fashioned mechanical models of the Universe are being trans-formed by modern subatomic physics, a development that has profound spiritual and philosophical ramifications affecting us all.

'Zohar's theory is brilliantly topical, answering the environment-alists' prayers for a holistic approach to nature and an escape from the spirit/matter dualism that has dominated western thinking from Plato to the modern Christian church.'   *Guardian*

'Zohar takes Fritjof Capra's classic *The Tao of Physics* several steps further in this provocative use of quantum theory . . . A stimuating work, thoughtfully conceived and unusually thorough.'   *Kirkus*

'Danah Zohar . . . successfully integrates modern physics not only with consciousness but also with the individuality of the human being within the context of society and the cosmos. I recommend *The Quantum Self* highly . . .'

Professor David Bohm, Emeritus Professor of Theoretical
Physics, Birkbeck College, London

■ *f l a m i n g o*

Bart Kosko

Fuzzy Thinking

The New Science of Fuzzy Logic

'An exciting and truly revolutionary book'
Danah Zohar, *Independent on Sunday*

For more than 2000 years, Western science has been based on absolutes. Things are black or white, alive or dead, all or nothing. As human beings we know the world is not really like this, that degrees exist between the extremes. But until now science has been unable to accommodate these uncertainties. Fuzzy logic is a scientific revolution that has been waiting to happen for decades – and its central tenets will dramatically change the relationship human beings have with the world. The question is to what degree.

In this absorbing, iconoclastic account of the head-spinning possibilities for fuzzy technology, Bart Kosko, fuzzy logic's most famous and combative apostle, urges us to abandon the debilitating binary world and turn to the East, for the future will be *fuzzy*.

'*Fuzzy Thinking* is about . . . a radically different way of structuring our thoughts and experience . . . that transforms our perception of reality'   Danah Zohar, *Independent on Sunday*

'Bart Kosko is the quintessential scientific cyberpunk – a hip, street-smart prophet of life in the Information Age'
Los Angeles Times

ISBN 0 00 654713 3

A Darwin Selection

Edited by Mark Ridley

Charles Darwin, almost uniquely among great scientists, wrote for the general public. For this *Darwin Selection*, Mark Ridley has chosen the key passages from Darwin's nine most important books, and for each of them he has filled in the context of the selection, annotated the few obscure points, and drawn on the latest Darwin scholarship to explain their history.

From the *Origin of the Species*, we have Darwin's beautifully clear exposition of natural selection and of the case against creationism; fom the *Descent of Man* we have his explanation of human intelligence and morality, and his theory of sex differences; and from *Coral Reefs* we have his theory of the origin of coral atolls – a theory that is still widely accepted today. We see him as an experimentalist, unveiling the loves of the plants; as a travel writer describing 'that little world within itself', the Galapagos Islands; and as a natural philosopher, serenely calculating how the actions of worms over long periods emerge as a geological force and the agency of archaeological preservation. *A Darwin Selection* contains many memorable details too; we can rediscover, for instance, the rudimentary tip of the human ear – the curiosity that finally introduced evolution to the polite conversation of the Victorian sitting-room . . .

ISBN 0 00 686321 3

The Fontana History of Astronomy and Cosmology

John North

Astronomy is not only a field of scientific research capable of the most startling discoveries; it is the oldest of the exact sciences. It is fitting then that in this lucid, elegant book, Professor North devotes particular attention to both the earliest and the most recent developments.

Stressing the indispensibility of an understanding of the heavens for the elementary calendrical calculations vital in all societies, Professor North demonstrates how surveying the skies helped generate the geometrical and mathematical principles crucial to early science in the Middle East and Greece, which in turn continued to underpin advances in astronomy right through the revolutions in thinking achieved by Copernicus, Kepler and Newton.

Astronomy, North shows, has a history marked by continuity. It offers a powerful illustration of concentrated, progressive scientific endeavour and yet it has also been a complicated and many-sided enterprise, integrating metaphysical, religious and cosmological speculations with the down-to-earth practical skills needed, for example, for navigation and timekeeping. In assessing the social position of astronomers, Professor North deftly explores the tensions between these different roles.

0-00-686177-6

Fontana History of Science
Series Editor: Roy Porter

The Fontana History of Chemistry

William H. Brock

The Fontana History of Chemistry, which draws extensively on both the author's own original research and that of other scholars worldwide, is conceived as a work of synthesis for the 1990s. Nothing like it has been attempted in decades. Beginning with the first tentative chemical explorations where primitive technology and techniques were deployed, Dr Brock proceeds via the alchemists' futile, but frequently profitable, efforts to turn lead into gold to recount the emergence of the modern discipline of chemistry as fashioned by Boyle, Lavoisier and Dalton. He provides a particularly generous critical emphasis on the roles of purity, analysis and synthesis, the exploration of reaction mechanisms, and the industrialization of chemical change, while also weighing up how chemistry has been taught and disseminated as a subject and as a science.

Although the goal of immortality was never achieved by the alchemists, history itself can act as a philosopher's stone, enabling us to absorb the experiences and ideas of those who lived before us. While brilliantly successful in explaining and exploiting chemical change, modern chemistry – in academy and factory alike – with its recondite language and symbolism and its associations with pollution and danger, prompts more fear than excitement in the uninitiated bystander. This book seeks to enlighten that bystander: to assess links between theory and practice, to reveal recurrent cycles and themes, and to encourage a heightened awareness of the human dimensions of the chemical sciences which might in turn promote a better public under standing of chemistry and the modern chemical and pharmaceutical industries.

ISBN 0 00 686173 3

In the Blood

God, Genes and Destiny

Steve Jones

'Steve Jones is one of the best storytellers around today.'
Independent on Sunday

'Few scientists write well for a general audience, but Steve Jones is exceptional.'
Observer

Genetics is at the heart of modern science. Genes link the past with the present and contain within themselves the fate of many of those who carry them. Every week sees new and startling discoveries in this, the most wonderful of sciences. In this book Professor Steve Jones asks what really is 'in the blood' – and just what that means.

In the Blood shows how genetics is coming uncomfortably close to the questions asked by philosophy, theology and even politics. It deals with issues of fate, of life and of death. To some, science threatens human autonomy. If everything is in the genes, what is left for free will? If man is but a glorified ape, where is the soul? Indeed, if society is just a mechanism for ensuring that genes are transmitted, what room is there for good and evil? If anyone is fit to begin to pass judgement on these fundamental questions – each one an ancient problem newly phrased in the language of science – it is Steve Jones, who is that rarest of animals, a scientist who is sceptical about his science. He asks these questions in a novel way and, sometimes, gives the answers.

ISBN 0 00 255511 5